Thorsten Rahn

**Heterotic Target Space Dualities with cohomCalg**

Thorsten Rahn

# Heterotic Target Space Dualities with cohomCalg

## An analysis using line bundle cohomology and cohomCalg++ Koszul

Südwestdeutscher Verlag für Hochschulschriften

**Impressum / Imprint**
Bibliografische Information der Deutschen Nationalbibliothek: Die Deutsche Nationalbibliothek verzeichnet diese Publikation in der Deutschen Nationalbibliografie; detaillierte bibliografische Daten sind im Internet über http://dnb.d-nb.de abrufbar.
Alle in diesem Buch genannten Marken und Produktnamen unterliegen warenzeichen-, marken- oder patentrechtlichem Schutz bzw. sind Warenzeichen oder eingetragene Warenzeichen der jeweiligen Inhaber. Die Wiedergabe von Marken, Produktnamen, Gebrauchsnamen, Handelsnamen, Warenbezeichnungen u.s.w. in diesem Werk berechtigt auch ohne besondere Kennzeichnung nicht zu der Annahme, dass solche Namen im Sinne der Warenzeichen- und Markenschutzgesetzgebung als frei zu betrachten wären und daher von jedermann benutzt werden dürften.

Bibliographic information published by the Deutsche Nationalbibliothek: The Deutsche Nationalbibliothek lists this publication in the Deutsche Nationalbibliografie; detailed bibliographic data are available in the Internet at http://dnb.d-nb.de.
Any brand names and product names mentioned in this book are subject to trademark, brand or patent protection and are trademarks or registered trademarks of their respective holders. The use of brand names, product names, common names, trade names, product descriptions etc. even without a particular marking in this works is in no way to be construed to mean that such names may be regarded as unrestricted in respect of trademark and brand protection legislation and could thus be used by anyone.

Coverbild / Cover image: www.ingimage.com

Verlag / Publisher:
Südwestdeutscher Verlag für Hochschulschriften
ist ein Imprint der / is a trademark of
AV Akademikerverlag GmbH & Co. KG
Heinrich-Böcking-Str. 6-8, 66121 Saarbrücken, Deutschland / Germany
Email: info@svh-verlag.de

Herstellung: siehe letzte Seite /
Printed at: see last page
ISBN: 978-3-8381-3393-5

Zugl. / Approved by: München, LMU, Diss, 2012

Copyright © 2012 AV Akademikerverlag GmbH & Co. KG
Alle Rechte vorbehalten. / All rights reserved. Saarbrücken 2012

To my siblings

# Acknowledgments

The research content of this dissertation was developed during the past two years and many people supported me during this time in various ways.

First of all I would like to thank my supervisor Ralph Blumenhagen for his continuous guidance and providing me with deep insights and physical as well as mathematical intuition. Furthermore I would like to thank Dieter Lüst for giving me the opportunity to broaden my knowledge on various summer and winter schools as well as for advice on technical questions and on questions concerning public and scientific talks. I am also deeply grateful to Stefan Groot Nibbelink for repeatedly long and enlightening discussions and for comments on the manuscript.

For scientific discussions on various physical topics I would also like to thank Thomas Grimm, Stephan Stieberger, Ilarion Melnikov, Bernhard Wurm, James Gray, Lara Anderson and Eric Sharpe. For pure mathematical discussions as well as for pointing out mathematical references I would like to thank Volker Braun, Hal Schenck, Rhys Davies, Phillip Candelas and Ron Donagi.

Next I would like to thank my collaborators and former office mates Benjamin Jurke and Helmut Roshy whose great competence in programming as well as pure math gave me the chance to learn a lot. With much pleasure I am looking back to many nice office days with them as well to having burgers at the Irish pub.

Special thanks goes to Oliver Schlotterer for sharing a great time in Munich, for deep and enlightening discussions on various jogging tours as well as for detailed feedback on the manuscript. I would also like to thank Frederik Beaujean, Peter Patalong and Christian Römelsberger for discussions and challenging climbing trips.

For a good atmosphere at the institute, for comments, discussions as well as for a lot of fun at social activities I would like to thank Andreas Deser, Felix Rennecke, Sebastian Halter, Sebastian Moster, Noppadol Mekareeya, Stefan Hohenegger, Johannes Held, Patrick Kerner, Federico Bonetti, Raffaele Savelli, Xin Gao, Pramod Shukla, Matthias Weissenbacher, Alois Kabelschacht and many others.

Last but not least I would like to express my deep gratitude to my siblings and my parents who always supported me in any way they could during the past years. I would furthermore like to thank Ricarda Marhauer for support before and during my first long public talk and Roland Marhauer for instructive tutorials on AS as well as MW2.

# Contents

Acknowledgments   2

Contents   6

1   Introduction   7

2   Heterotic String Compactifications   19
    2.1   Spacetime description of the heterotic string . . . . . . . . . . . . 20
    2.2   The massless spectrum . . . . . . . . . . . . . . . . . . . . . . . . 23
    2.3   Worldsheet description of the heterotic string & GLSMs . . . . . 27

3   Toric Geometry   39
    3.1   Physical motivation: Gauged linear sigma models and flux compactifications . . . . . . . . . . . . . . . . . . . . . . . . . . . . . . 39
    3.2   From projective spaces to toric varieties . . . . . . . . . . . . . . . 40
    3.3   Lattice polytopes and fans . . . . . . . . . . . . . . . . . . . . . . 45
    3.4   Line bundles and divisors . . . . . . . . . . . . . . . . . . . . . . . 51

4   Cohomology of Line Bundles   55
    4.1   Physical motivation: Heterotic GUTs, line bundles on D7 branes and type IIB/F-theory instantons . . . . . . . . . . . . . . . . . . 55
    4.2   Cohomology: The idea . . . . . . . . . . . . . . . . . . . . . . . . 56
    4.3   Sheaf cohomology . . . . . . . . . . . . . . . . . . . . . . . . . . . 60
    4.4   Sheaf-module correspondence . . . . . . . . . . . . . . . . . . . . 70
    4.5   Commutative algebra . . . . . . . . . . . . . . . . . . . . . . . . . 72
    4.6   Cohomology of line bundles: The algorithm . . . . . . . . . . . . 85
    4.7   Explicit computations . . . . . . . . . . . . . . . . . . . . . . . . 89

5   Equivariant Cohomology   99
    5.1   Physical motivation: Orientifolds in type IIA/B and heterotic orbifolds . . . . . . . . . . . . . . . . . . . . . . . . . . . . . . . . . . 100
    5.2   Topological invariants for $\mathbb{Z}_2$ involutions . . . . . . . . . . . . . 100
    5.3   An algorithm conjecture for $\mathbb{Z}_2$-equivariance . . . . . . . . . . . 103

|     |     |     |
| --- | --- | --- |
| 5.4 | Invariants for finite group actions | 107 |
| 5.5 | Some explicit examples for finite group equivariance | 108 |
| 5.6 | Generalized equivariant algorithm conjecture | 116 |

## 6 Subvarieties and Calabi-Yau Manifolds — 119

- 6.1 Physical motivation: Calabi-Yau compactifications, D-branes and GUT divisors in F-theory ... 119
- 6.2 Calabi-Yau spaces ... 120
- 6.3 Line bundles and their cohomology ... 121

## 7 Vector Bundle-Valued Cohomology — 129

- 7.1 Physical motivation: Heterotic GUTs, moduli spaces of heterotic and type IIA/B theories ... 129
- 7.2 Bundle-valued cohomology ... 130
- 7.3 Tangent bundle-valued cohomology ... 132
- 7.4 Vector bundle-valued cohomology ... 142

## 8 Purely Combinatorial Approach to Cohomology — 153

- 8.1 Lattice polytopes and Calabi-Yau hypersurfaces ... 153
- 8.2 Cayley polytopes and CICYs ... 156

## 9 Target Space Dualities in the String Landscape — 165

- 9.1 The setup and assumptions ... 166
- 9.2 Explicit construction of dual (0,2) models ... 169
- 9.3 Landscape studies ... 190

## 10 Conclusions and Outlook — 199

- 10.1 Cohomology of line bundles ... 199
- 10.2 Equivariant cohomology ... 201
- 10.3 Combinatorial cohomology ... 201
- 10.4 Targe space dualities ... 202

## A Anomaly Cancellation — 205

- A.1 Anomaly cancellation for target space dualities ... 205

List of Figures — 209

List of Tables — 211

Bibliography — 212

# Chapter 1

# Introduction

We are living in exciting times for particle physics. The large hadron collider (LHC) at CERN in Geneva has been running for a while now and has reconfirmed almost every detail we know about the *standard model of particle physics* (SM) at energy scales beyond those tested in the past. Still one quite important question which remained unanswered within the SM is the existence of the Higgs Boson whose detection is top priority of the experiment and might be answered soon. At the moment, both experiments, ATLAS as well as CMS, find signals that might have been caused by the Higgs particle but are yet to be validated. The SM describes the three particle interactions, i.e. electromagnetic, weak and strong interaction via quantum field theories with (non-)Abelian gauge symmetries [1–3] and is based on the full gauge group $SU(3) \times SU(2) \times U(1)$. It successfully describes the first two of these interactions in a combined framework, called electroweak interaction, with gauge group $SU(2) \times U(1)$ containing the leptons, i.e. electrons, muons, taus and their neutrinos. On the other hand, the quarks align into multiplets of $SU(3)$ color symmetry. Based on these concepts, in the past various SM predictions have been confirmed by experiments.

Potential issues and extensions of the standard model

Though the SM is very successful in many ways there are also quite some problems as well as open questions coming with it.

*Quantum gravity.* The first and most important issue of the SM is that though it describes three of the fundamental interactions, it lacks the description of gravity within this framework. The attempt to include gravity on the same footing as other interactions leads to a non-renormalizable theory. Hence, if one wants to get a sensible theory, unifying all the four interactions, one has to come up with a different approach.

*Dark matter and energy.* Cosmological observations as the orbital velocities of stars inside galaxies lead to contradictions with the theory and in order to overcome these one has to introduce a new kind of matter and energy that is not visible to us. It is, hence, called dark matter and dark energy and they make up 74 % and 22% of the overall amount respectively while the visible part of the universe covers less than 5 %. The SM has no explanation for both dark matter and dark energy.

*Cosmological constant.* Einsteins equations allow for an extra term that is proportional to the metric. Its coefficient is called the cosmological constant, whose value basically describes the energy of the vacuum and the prediction by the SM differs from measurements by 120 orders of magnitude [4]. This does not imply that the value is small since the universe contains a fair amount of dark energy as we stated above. The mismatch in the value of that constant, however, is one of the biggest puzzles in theoretical physics.

*Grand unification.* The SM is no unified theory of strong and electroweak interactions. In order to unify them one has to introduce a gauge group that contains the SM group as a subgroup and whose full symmetry is broken at low energies. Such theories are referred to as grand unified theories (GUTs) and may come with an $SU(5)$ [5] or $SO(10)$ [6] gauge group.[1]

*Parameters.* Another issue is that there is some sort of arbitrariness in the SM since there are around twenty independent parameters that describe the Fermion masses, the mixing angles, the gauge couplings as well as the Higgs mass and the Higgs self-coupling. These parameters cannot be predicted by the SM but have to be measured and put into the theory by hand. Furthermore, the original version of the SM predicted the neutrinos to be massless. The measurement of neutrino oscillations requires a modification of the SM that contains massive neutrinos. Also, their masses as well as mixing angles enter as free parameters that cannot be predicted.

*Open questions.* There are a couple more question that are not answered by the SM itself, for instance, why are there three generations of Fermions, why are we living in four dimensions, why is the low-energy gauge group $SU(3) \times SU(2) \times U(1)$, why is there a confinement of the quarks in hadrons and why is the weak force so much stronger than gravity (hierarchy problem), just to name some.

---

[1] In Georgi-Glashows model [5] the GUT group $SU(5)$ is used to accommodate the SM gauge group $SU(3) \times SU(2) \times U(1)$. In this model the quarks and leptons are arranged in the representations $\bar{5} \oplus 10 \oplus 1$ of $SU(5)$. Nevertheless at least this particular model has been excluded since it predicts a decay of the proton that could be disproved by experiments.

## Supersymmetry

One possible extension of the SM is the introduction of a symmetry that relates Bosons and Fermions, so-called *supersymmetry* (SUSY) (see e.g. [7–9]). It was shown by Coleman and Mandula [10] that under certain mild conditions on the $S$-matrix of a quantum field theory there is no way to produce any theory with a symmetry group different from the direct product of the Lorentz group and the internal gauge group. This is the first thing that SUSY is able to bypass via introducing a set of Grassman odd generators forming the generalized version of a Lie algebra called super algebra. SUSY relates Bosons with Fermions that are transforming in the same group representation and since the SM does not provide any such multiplet, all the SM fields in a supersymmetric SM will obtain a superpartner. Here to each Boson is added a new Fermion and to each Fermion a new Boson. The most studied supersymmetric extension of the SM is called the *minimal supersymmetric standard model* (MSSM) [11] that realizes $\mathcal{N} = 1$ SUSY. It contains a vector multiplet that accommodates the SM gauge fields as well as their superpartners, called gauginos and a chiral multiplet that contains all the Fermions, i.e. quarks and leptons along with their superpartners called squarks and sleptons, as well as the Higgs and its superpartners, the Higgsinos. Moreover, if SUSY is present at high energies, it has to be broken at some energy above the observed scales, otherwise the particles supersymmetric to the SM particles would have the same mass as the SM particles and we would have observed them by now. SUSY breaking allows them to be heavier at low energies. The MSSM has several nice features:

*Hierarchy problem.* If the SUSY braking scale is in the TeV range which means that the superpartners of the SM are not significantly heavier than the Higgs, one can see that the Higgs mass is not only corrected by Bosonic loop diagrams, but also by Fermionic loop diagrams with the opposite sign. This could explain why the Higgs mass is so much lighter than the Planck mass and is a possible solution of the aforementioned hierarchy problem.

*Gauge coupling unification.* A second nice feature of the MSSM is that, assuming again the SUSY breaking scale is near the TeV scale, the three gauge coupling constants are unified at high energies [12] while they miss without SUSY.

*Dark matter.* The third issue that can be resolved by such a scenario is the description of dark matter. The lightest superparticle of the MSSM is stable and massive, assuming $R$-parity is conserved. If it is furthermore neither strongly nor electromagnetically interacting, it can provide a candidate for dark matter.

Although the MSSM seems to solve a lot of problems, one major drawback of such a theory is that the space of free parameters increases to about one hundred and the question arises where the values of these parameters have their origin. One should also mention that recently an extensive search for SUSY particles was performed at the LHC and so far no evidence for their existence has been found. But even though some parts of the parameter space of MSSMs have been scanned, there is still room left and so far the MSSM is not yet excluded. However, even if low-energy SUSY is not realized by nature, many nice properties still hold for high-energy SUSY. What cannot be addressed in this scenario, though, is the solution of the hierarchy problem without fine tuning.

Besides the MSSM, there are also other possible extensions with SUSY and the most canonical one right after the MSSM is the NMSSM (next to MSSM). Here additional chiral superfields appearing as singlets are present [13–16].

In theories with non-Abelian symmetries, promoting them to local symmetries i.e. gauge symmetries provides interaction terms. So it seems reasonable also to turn on local SUSY. This implies diffeomorphism invariance which allows for the inclusion of a spin two field to the theory. Every quantum theory of gravity has to contain such a field that represents the exchange particle of the gravitational interaction and is referred to as the graviton. In a supersymmetric framework it is possible to include the graviton along with its Fermionic superpartners called gravitini which carry spin $3/2$ in a supermultiplet. Such a supersymmmetric graviational theory is called supergravity (SUGRA) and can be seen as low-energy effective field theory of some UV completed theory.

## String theory

Every quantum field theory including the SM contains a very fundamental assumption. Namely they consider the smallest object to be zero-dimensional, hence a point. Assuming instead the fundamental object to be one-dimensional, i.e. a string, yields a new kind of theory, called *string theory*. Consequently in string theory the one-dimensional world line of the point particle will be replaced by a two-dimensional surface, called the *worldsheet*. Going from point to string is of course compatible with our observations as long as the string is small enough. If this is the case it cannot be distinguished from a point particle and in fact, besides the argument of simplicity, there is no reason to favor the "point assumption" over the "string assumption". One might be worried about the fact that this change destroys locality but it turns out that this even results in the advantage that particular Feynman diagrams remain finite. Some relevant features of string theory for us are the following:

String theory naturally carries the graviton, the spin two quantum of the gravitational interaction, furthermore can be reduced to general relativity in a low-energy limit and therefore describes a quantum theory of gravity. The amazing thing is that the only requirement to obtain general relativity is the seemingly completely unrelated invariance of the world sheet under conformal, i.e. angle preserving, transformations.

SUSY is not only natural but even crucial to a string theory that shall admit for Bosons as well as for Fermions. The former ones are called Bosonic string theories while the latter are referred to as superstring theories. Hence superstring theories can provide all the nice features of supersymmetic theories and furthermore, allow for a derivation of the low-energy parameters for a given string vacuum.[2] Then the only remaining free input parameter of the theory is the length of the string or equivalently its tension. Consequently since string theory combines SUSY and gravity it provides a UV completion of low-energy SUGRA. Furthermore, in general there appear conformal anomalies on the worldsheet and in order to cancel them one has to impose a certain dimension of our space time which means that string theory predicts its own dimensionality.

Nevertheless one should mention that so far no consistent version of a string theory that completely reproduces all the features of the SM is known and therefore is yet to be found. Here, even though there are no conceptual problems, there are technical diïñČculties that so far obstructed the construction of such a model. One of the major challenges is to avoid the appearance of additional massless fields, called moduli, that are not observed by experiments.

As already mentioned, a Bosonic string theory is not suited to describe nature since it will only allow for space time Bosons. Moreover it also predicts tachyonic states, violating causality. In such a scenario we would find the critical dimension of the string to be 26. Superstring theories on the other hand do allow for space time Fermions and M. Green and J. Schwarz [17] could show that if one chooses the critical space time dimension to be ten, there are no quantum anomalies and one can get rid of all the tachyons [18]. Furthermore, there is at first sight not a unique way to implement a superstring and one can see that there are five different ways to do so. These five string theories are called *type I*, *type IIA*, *type*

---

[2]We will see below how the string vacuum can be identified with a geometric space of very special properties. In the early years it was hoped that there might be a unique choice for such a space but in the last years it turned out that this choice is far from unique. There is a large number of geometries to choose from and one might reason that the choice of the space is no better than the choice of the parameters. On the other hand the set of geometries is at least discrete and probably even finite. Furthermore, all the parameters arising from it have a very nice geometrical meaning which is also an esthetical improvement and moreover there might still be some reason behind the particular choice and maybe even a mechanism exists that singles out the right one.

*IIB, heterotic SO*(32) and *heterotic* $E_8 \times E_8$ that finally turn out to be related via string dualities and therefore are basically five different viewpoints of the same thing [19]. They all differ in some way, for instance in the type II theories we have open strings as well as closed strings. Here the open strings end on higher dimensional objects referred to as D-branes and they describe the gauge interactions of the theory whereas the closed strings give rise to the gravitational interactions. On the other hand heterotic string theories only contain closed strings and the gauge fields arise due to the presence of a so-called gauge bundle that comes along with the ten dimensional space time. Such a gauge bundle is basically like a vector bundle, i.e. a construction that attaches a vector space to every point in space time and the gauge fields here arise as the so-called connection of the gauge bundle which roughly speaking tells us how to connect the attached spaces at nearby points. Furthermore, there is an eleven-dimensional theory called *M-theory* that also can be connected to the five superstring theories whereas itself does not have strings as fundamental objects but two-dimensional membranes. It can moreover be considered to be the limit of type IIA where the string coupling is large. Then using string dualities all five string theories can be recovered starting from M-theory and, hence, it is a more general framework that reduces to string theory in certain limits.

## GUTs in string theory

From now on we will be concerned mostly with heterotic string theories in which the aforementioned GUTs arise quite naturally. One of the advantages of $E_8 \times E_8$ heterotic string theories is that they naturally come with an exceptional gauge group that can be broken down to some GUT group $E_6$, $SO(10)$ or $SU(5)$ in the four-dimensional theory. This broken gauge group can then be used to reproduce the SM at some lower scale and in contrast to the model of Georgi-Glashow, which was ruled out, the $SU(5)$ model in this context may still be realized by nature. In fact MSSM-like models have been constructed explicitly within the context of heterotic strings [20–27], in the type II setting using intersecting D-branes (for a review see [28]) as well as in various other scenarios [29–41].

One advantage comparing with the plain MSSM is that in principle all the information of the model, e.g. the parameters of a string-based model may be obtained for instance by the geometry of the extra dimension which for the heterotic case works as follows: We already mentioned the critical dimension of the heterotic string to be ten and if we consider this ten-dimensional space time to split into a direct product of four-dimensional Minkowski space with a six-dimensional space $\mathbb{R}^{3,1} \times \mathcal{M}$, we have to choose these six extra dimensions small, in order to meet the experimental observation of a four-dimensional universe. This procedure is referred to as *compactification*. Sure enough the size of $\mathcal{M}$ has to be chosen below the length scale observed in experiments. Considering wave equations on a ten-dimensional space $\mathbb{R}^{3,1} \times \mathcal{M}$ one can see that e.g. massless ten-dimensional fields will give rise to additional massive and massless fields in four dimensions corresponding to those with momentum components in the compact directions. Their masses usually scale with the inverse "radius" of the compact manifold. Hence, in the limit where $\mathcal{M}$ is small these masses become too large to be detected and only the massless modes will survive. The former ones are called Kaluza-Klein modes after T. Kaluza and O. Klein who first considered such a compactification scenario almost a hundred years ago, where they tried to unify electromagnetism and general relativity by introducing a compact fifth dimension. The number of zero modes we obtain from a given compact geometry at the end depends on the topology of that manifold.

In the heterotic setting we can find these zero modes as follows. As long as the string length is much smaller than the size of the internal compactification space, we can give an effective description of string theory in terms of ten-dimensional SUGRA which is also called heterotic SUGRA. Here the internal dimensions of the heterotic string form indeed a smooth manifold $\mathcal{M}$ together with a gauge bundle and it is possible to derive the properties of the four-dimensional theory such as the GUT group, the spectrum, Yukawa couplings etc. only from the topology of the six-dimensional manifold. In particular the chiral spectrum can be derived by computing the zero modes of the Dirac operator. Only these zero modes will survive the Kaluza-Klein compactification process and in fact one can state a relation between such zero modes and certain cycles of the compact space. For instance a torus has two different one-dimensional cycles corresponding to the two circles that span the torus. The set of independent cycles is referred to as *homology* of the manifold whereas its dual version is called the *cohomology* and contains differential forms. Hence, the task of finding the surviving zero modes will be equivalent to finding particular differential forms defined on $\mathcal{M}$ which are completely determined by its topology. So at the end in order to understand the physics of the models we build we will not be able to avoid the investigation of the topological data of $\mathcal{M}$. In fact we will find some topological constraints that we can impose more easily than others. For instance preserving SUSY requires a compactification space that allows for a nowhere vanishing covariantly constant spinor. This property is known to mathematicians that called such spaces Calabi-Yau spaces after E. Calabi and S.T. Yau. Studying Calabi-Yau spaces in detail is, hence, crucial to model building in string theory.

## String model building with toric geometry

Usually Calabi-Yau manifolds are quite abstract and it is very hard to obtain any information about them. But there is one mathematical framework in which Calabi-Yaus are accessible and this is where they are realized as sub-spaces of so-called toric varieties. Pedagogical introductions for physicists can be found e.g. in [42,43] and more advanced ones in [44–46]. In this setting the requirement of $\mathcal{M}$ to be Calabi-Yau is rather easy to meet but the phenomenological requirement to give the right number of generations, for instance, is highly non-trivial and needs explicit calculations. A realistic model will need quite a variety of particular phenomenological properties. These are realized as topological properties of the corresponding geometry and have subtle connections. Thus changing the geometry in favor of some of the requirements might also change others and one has to choose the geometry such that all the requirements are fulfilled simultaneously. In other words, it is quite involved to engineer such a model from scratch. Therefore many attempts are to take a given set of geometries and scan them for the right configuration or at least to learn more about the problems that come with the process [25–27, 47–57]. Howsoever one might approach the matter of model building, at the end there is no way to get around the task of calculating topological quantities of the compactification space. If we work with toric varieties one can see that these topological quantities are all encoded in the topology of holomorphic vector bundles of rank one, i.e. *holomorphic line bundles* which are spaces constructed by attaching a complex space $\mathbb{C}$ to each base point. Hence, the task one is left with is the calculation of cohomology groups of line bundles which basically represent the topological structure of the line bundles.

So we will need efficient ways to calculate line bundle cohomology in order to handle the physics behind these models. For the case that the Calabi-Yau space is given as a transverse intersection of hypersurfaces in products of projective spaces, which are the very simplest set of toric varieties, there exist index formulæ. Thus there is no need to calculate the line bundle cohomology precisely since one can take the shortcut via the corresponding index. Even though this opens a variety of possibilities to build string models, these spaces are still fairly constraint and do not carry all the subtle structure of generic toric varieties. Hence, certain features that are present for the generic case will never be present in those more simple cases and at the end we might not be able to find the correct compactification space within this set of geometries. For these reasons it is necessary to go beyond the most elementary set of products of projective spaces which raises the task of calculating line bundle cohomology properly.

## Content of this book

This brings us to the actual content of this book. The unavoidable task of calculating line bundle cohomology on toric varieties requires efficient tools since it has to be performed possibly hundreds of times for single models. There are a lot of tools around to calculate various things within the framework of toric geometry [58–65] but so far no sufficiently efficient algorithm existed which is why we started working on this. We proposed a conjecture along with a Wolfram Mathematica implementation of such an algorithm [66] which was later on proven by our group [67] and independently by the UPenn math department [68]. Later we also provided an even faster C++ implementation of the algorithm which we called **cohomCalg** and in addition a Wolfram Mathematica interface that also contains the routines to obtain the cohomology of higher-rank vector bundles restricted to the intersection of hypersurfaces in the toric variety, named **cohomCalg Koszul** extension. The full package including a very explicit manual is available online [69]. Furthermore, we investigated generalizations of this algorithm to equivariant cohomology [70, 71] that makes it possible to consider a further discrete action on the base manifold which in string theory are called orbifolds [72–75]. There we also investigated the relation of line bundle cohomology to the different contributions of the formula of Batyrev [76]. This formula makes use of a purely combinatorial approach to calculate the Hodge numbers of a Calabi-Yau space and relates certain line bundle cohomology groups to the so-called twisted sector of the corresponding Landau-Ginzburg orbifold theories. These sectors are special to orbifold constructions and arise due to the identification via the discrete action. Finally the algorithm also found its application in the investigation of so-called *target space dualities* that relate different heterotic GUT models with one another. This shed light on many different aspects of the duality and it was tested in the scan of a large landscape of models [77, 78] where **cohomCalg** had to be used tens of millions of times.

## Outline

The content of this book is organized as follows:

*Chapter 2:* We will first introduce some basics on the heterotic string and model building in that context. We will see how the physical spectrum is related to the geometry of the internal dimensions and provide the requisites needed in chapter 9. Furthermore, we will see that this motivates the need to study the insights to the theory of toric varieties and line bundles.

*Chapter 3:* In this chapter we will give a very basic introduction to toric varieties that were motivated by chapter 2. Here toric varieties will be built up as generalizations of projective spaces and we will try to give an intuition in addition to the abstract definitions. Especially we will focus on the computational side of it and try to provide the reader with everything that is needed for model building in string theory. Furthermore, we also explain the concepts of divisors and line bundles on toric varieties.

*Chapter 4:* Next we are going to introduce the notion of cohomology and in particular the cohomology of line bundles. The whole chapter is devoted to the task of deriving and proving the algorithm for the calculation of line bundle cohomology, finally stated in theorem 4.6.1. It will therefore cover the content of [66, 67] and will also comment on [69].

*Chapter 5:* One nice feature of the algorithm is that it provides explicit representatives of the cohomology groups. Therefore it is possible to see how a discrete action affects these representatives. The explicit procedure to obtain the equivariant cohomology of such a discrete action is explained in this chapter. We conclude with a conjecture which is based on the observation comparing our results with the Lefschetz theorem which is an index theorem for such equivariant cohomologies. It can be found in its original version among other things in [70].

*Chapter 6:* Since we are also interested in the cohomology of line bundles that live on a subvariety of a toric variety, e.g. a Calabi-Yau manifold, we have to understand how one can get these cohomology groups from calculations on the ambient space. The way to do it is well known and makes use of the Koszul resolution. We will show how one can break this down to a set of short exact sequences that allow us to perform the calculations. Several examples are provided in order to demonstrate the techniques to those that are not familiar with them and may be skipped by the experienced reader. The extended implementation **cohomCalg Koszul** extension [69] is making excessive use of these methods.

*Chapter 7:* The next step is to use holomorphic line bundles (rank one) and to build up vector bundles of higher-rank (rank three, four or five). It turns out that also for this matter one can employ short exact sequences to boil the calculation of vector bundle-valued cohomology down to the case of line bundle cohomology. The monad vector bundles mentioned in chapter 2 that are used to find the chiral spectrum of a heterotic model are introduced and all the techniques that are crucial to the physical analysis in chapter 9 are described and demonstrated. Again **cohomCalg Koszul** extension [69] is making use precisely of these methods and it is worth to have a look at it before using the program blindly.

*Chapter 8:* This chapter is a little bit off the scope of the rest of the book. There are formulæ from Batyrev and Borisov [76,79] that allow one to use a purely combinatorial approach to the calculation of the Hodge numbers of Calabi-Yau manifolds realized as subvarieties in toric varieties. They provided a formula to calculate the so-called stringy Hodge numbers defined on potentially singular Calabi-Yau spaces that are basically coinciding with the Hodge numbers of a smooth crepant resolution in case that it exists. Using our algorithm we found that actually particular contributions to these numbers can be traced back to certain cohomology groups of line bundles on the ambient space [70]. In cases where a Landau-Ginzburg phase exists, these contributions correspond to states in the twisted sector and, hence, we can assign to the cohomological degree a particular physical meaning. Furthermore, we extend these ideas to codimension two cases and as a byproduct also provide an explicit combinatorial formula for Hodge numbers of a Calabi-Yau four-fold that may become useful in the F-theory setting.

*Chapter 9:* Finally in the last chapter we apply our algorithm to construct heterotic models. In particular we investigate target space dualities of heterotic models with $(0, 2)$ worldsheet SUSY. The fast computational abilities of **cohomCalg Koszul extension** allowed us to calculate many explicit examples and, hence, to give an explicit description of the construction of these models. In particular we focused on heterotic models where the holomorphic vector bundle is given by a deformation of the tangent bundle which is a rank three bundle [77]. We furthermore explain how one can employ conifold transitions to relate the base manifolds of the initial model to the base manifolds of the dual models. We also show how chains of dual models can be produced and provide data of a scan of a large landscape of models. This indeed provides strong evidence for the conjecture that target space duality is indeed a duality of the full string models. The more general situations of $E_6$, $SO(10)$ and $SU(5)$ GUTs where the holomorphic vector bundle is no longer a deformation of the tangent bundle [78] are considered, too.

# Chapter 2

# Heterotic String Compactifications

We have pointed out in the introduction that there are different types of string theories. For example, the Bosonic string which is forced to be defined in a 26-dimensional space time, due to anomaly cancellation conditions. On the other hand, consistency of string theories with space time supersymmetry such as the type IIA/IIB and type I require in a ten-dimensional environment. Both Bosonic as well as superstring theories involve independent left and right movers in the mode expansion of the closed string.

An idea that seems on first sight a little counter intuitive is to combine these two concepts to a so-called *heterotic* model firstly introduced by Gross et al. [80, 81]. This hybrid model merges left-moving modes of the Bosonic string and the right-moving modes of the superstring into one single theory. While the type II string theories have $\mathcal{N} = 2$ supersymmetry in ten dimensions, the heterotic string theory only preserves $\mathcal{N} = 1$ SUSY in ten dimensions. As in the type II setting, anomalies arise and they have to be canceled out which, as one can show, enforces the gauge group to be either $SO(32)$ or $E_8 \times E_8$. The standard model gauge group $SU(3) \times SU(2) \times U(1)$ can be embedded into $E_8$ and hence heterotic string theories provide a natural environment for GUT model building. The other $E_8$ can be considered invisible and this way does not influence the observable physics.

In string theory we usually have two different viewpoints of the same scenario in hand, the spacetime description and the worldsheet description of the string. The space time description is valid once the string scale is small compared to the size of the compactification space. If this assumption is not valid anymore, string effects apply and the geometric description may no longer be appropriate. In the following we will give a quick review for both, the spacetime and the worldsheet perspective of the heterotic string.

In both descriptions we will see how all the mathematical concepts, we are going to introduce in chapters 3 to 8, will beautifully arise and hence motivate the need to explore mathematical techniques in order to calculate quantities for real-world physics. Here the cross-fertilization between mathematics and physics manifests itself. In fact some of the mathematical concepts were motivated and even properly formulated because of the underlying physics and on the other hand the physicists made also extensive use of mathematics to understand the physics better. Taking advantage of these two perspectives is precisely what we did for the investigation of target space dualities as described in chapter 9.

## 2.1 Spacetime description of the heterotic string

For the spacetime description of the heterotic string we assume that all length scales defining our internal dimensions are large compared to the string length. Due to this assumption we don't have to deal with string effects and hence our string theory is effectively simply described by a field theory which is given by ten-dimensional SUGRA coupled to Yang-Mills theory. The massless string spectrum is then governed by an effective action

$$S = \int \mathrm{d}^{10}x \frac{(-G)^{1/2}}{\kappa^2 e^{2\Phi}} \left( R + 4|\partial\Phi|^2 - \frac{1}{2}|H_3|^2 - \frac{\alpha'}{4}\left\{\mathrm{tr}\left(|F_2|^2\right) - \mathrm{tr}\left(|R_2|^2\right)\right\}\right)$$
$$+ \text{Fermions}$$
(2.1)

which can be derived calculating for instance on-shell scattering amplitudes and contains the following fields:

$$\text{Dilaton } \Phi, \quad \text{10d metric } G, \quad B\text{-field } B_2, \quad \text{gauge field } A, \qquad (2.2)$$

and defined from those furthermore contains the three form flux $H_3$, the Riemann two-form $R_2$, the Ricci scalar $R$ and the gauge field strength $F_2$, corresponding to the gauge fields $A$. Furthermore the three-form $H_3$ can be written in terms of the $B$-field as

$$H_3 = \mathrm{d}B_2 + \frac{\alpha'}{4}\left(\mathrm{CS}_\omega - \mathrm{CS}_A\right), \qquad (2.3)$$

where $\mathrm{CS}_\omega$ and $\mathrm{CS}_A$ are the Lorentz and the Yang-Mills Chern-Simons forms of spin and gauge connection $\omega$ and $A$ respectively. They are given in terms of their connections in such a way that they obey the Bianchi identity:

$$\mathrm{d}H_3 = \frac{\alpha'}{4}\left(\mathrm{tr}\left\{R \wedge R\right\} - \mathrm{tr}\left\{F \wedge F\right\}\right). \qquad (2.4)$$

## 2.1. Spacetime description of the heterotic string

As one can see from (2.1), the gauge fields in the action are suppressed and come only into play at higher order in the string coupling. There are also Fermions in the above action (2.1), namely a gravitino $\Psi$ a dilatino $\lambda$ and a gaugino $\chi$. All the fields in the (2.1) accommodate a SUGRA multiplet and a vector multiplet where the fields $G$, $B_2$, $\Phi$, $\Psi$, $\lambda$ belong to the SUGRA multiplet while the $A$, $\chi$ belong to the vector multiplet and transform in the adjoint representation of the gauge group $E_8 \times E_8$ or $SO(32)$. In order to preserve supersymmetry, the variation of these Fermions have to vanish, i.e.

$$\begin{aligned}
\delta\Psi_C &= \left(\partial_C + \left(\frac{1}{4}\omega^{AB}{}_C - \frac{1}{8}H^{AB}{}_C\right)\Gamma_{AB}\right)\epsilon = 0, \\
\delta\lambda &= -\frac{1}{2\sqrt{2}}\left(\partial_A\Gamma^A\Phi - \frac{1}{12}H_{ABC}\Gamma^{ABC}\right)\epsilon = 0, \\
\delta\chi &= -\frac{1}{4}F_{AB}\Gamma^{AB}\epsilon = 0.
\end{aligned} \quad (2.5)$$

It turns out that in order to find solutions to the equations of motion of (2.1) which do actually preserve $\mathcal{N} = 1$ space time supersymmetry and furthermore $d = 4$ Poincare invariance, it is sufficient to solve the system of equations (2.5). Doing that we still need to satisfy the Bianchi identity given in (2.4) but after taking this into account we are done.

We are actually only interested in compactifications where globally no fluxes are present which means that in all our equations above we can formally put $H_3 = 0$. We also assume that our ten-dimensional space time splits into a flat four-dimensional one and a six-dimensional manifold

$$\mathcal{M}_{10} = \mathbb{R}^{1,3} \times \mathcal{M}. \quad (2.6)$$

Then the first SUSY variation in (2.5) simplifies to

$$\left(\partial_C + \frac{1}{4}\omega^{AB}{}_C\Gamma_{AB}\right)\epsilon =: \nabla_{\mathcal{M}}\epsilon = 0 \quad (2.7)$$

and reflects the fact that the spinor $\epsilon$ has to be covariantly constant on the internal six-dimensional manifold $\mathcal{M}$ that constraints the manifold to be a so-called Calabi-Yau manifold which we will introduce later in definition 6.2.1. The existence of such a spinor immediately forces the tangent bundle of the manifold $\mathcal{M}$ to have $SU(3)$ structure and it is also referred to as *Killing spinor*. Notice that the spinor $\epsilon$ lives in $SO(1,9)$ which is broken to $SO(1,3) \times SO(6)$ for a space of the form (2.6) and its six-dimensional part has right and left chiral pieces that transform under $SO(6) \cong SU(4)$ as **4** and **$\bar{4}$** respectively. Since the spinor of the space $\mathcal{M}$ gets rotated by elements of the holonomy group $\mathcal{H}$, we need to make sure that the **4** decomposition under $\mathcal{H}$ contains singlets. Otherwise we could not meet condition (2.7). This can be accomplished if we choose the holonomy group $\mathcal{H}$ to be $SU(3)$. Then we get the decomposition

$$4_{SU(4)} = \mathbf{3}_{SU(3)} + \mathbf{1}_{SU(3)} \tag{2.8}$$

and we can guarantee the existence of exactly one spinor satisfying (2.7).

After making sure that the gravitino variation is satisfied we still have to deal with the SUSY variations for the dilatino and the gaugino. The one for the dilatino is actually not to hard to accomplish. Since our flux $H_3$ is already formally put to zero, it suffices to just put the dilaton to a constant and we are done. However solving the equation for the gaugino in (2.5) we have to work harder. On first sight one might think that we could simply put $F_2$ to zero but this would contradict the Bianchi identity (2.4). This can be seen once we integrate on both sides over a four cycle, which gives zero on the left hand side:

$$\begin{aligned} 0 &= \int \mathrm{tr}\,(R \wedge R) - \int \mathrm{tr}\,(F \wedge F) \\ &\Leftrightarrow c_2\,(T_\mathcal{M}) = c_2\,(\mathcal{V})\,, \end{aligned} \tag{2.9}$$

where $\mathcal{V}$ is the vector bundle corresponding to the gauge connection. Since one can show that the second chern class of a Calabi-Yau manifold cannot vanish, it is clear that we cannot simply put $F_2$ to zero without violating (2.4). Furthermore we can constrain this non-vanishing field strength by the SUSY variation for the gaugino. This results into the constraints

$$F_{ij} = F_{\bar{i}\bar{j}} = 0 \tag{2.10}$$
$$F_{i\bar{j}}\, g^{i\bar{j}} = 0\,. \tag{2.11}$$

The first two equations (2.10) simply state that the bundle we choose has to be a holomorphic vector bundle and is actually not too hard to satisfy. The third equation (2.11) is called the specialized Hermitian Yang-Mills equation and it is much less trivial to solve for this equation. Our internal manifold $\mathcal{M}$ has to be a Calabi-Yau manifold and up to now there is not even one single Calabi-Yau known, where one can explicitly write down the Hermitian metric, even though its unique existence was proven in Yau's theorem [82] as stated in definition 6.2.1. Luckily there also is a similar theorem that proves the existence of a solution to the Hermitian Yang-Mills equations. It is due to Donaldson, Uhlenbeck and Yau [83, 84] and states the following:

**Theorem 2.1.1** (Donaldson-Uhlenbeck-Yau Theorem). *For a holomorphic vector bundle $\mathcal{V}$ there exists a solution to the hermition Yang-Mills equation*

$$F_{i\bar{j}}\, g^{i\bar{j}} = 0\,, \tag{2.12}$$

*iff $\mathcal{V}$ is $\mu$-stable. This means that the slope of every subsheaf $\mathcal{F} \subset \mathcal{V}$ is strictly smaller than the slope of $\mathcal{V}$ itself:*

$$\mu(\mathcal{F}) := \frac{1}{rk\mathcal{F}} \int_{\mathcal{M}} J \wedge J \wedge c_1(\mathcal{F}) < \mu(\mathcal{V})\,. \tag{2.13}$$

For $\mathcal{V}$, since its first Chern class vanishes, the slope of every subsheaf has actually to be negative. Since we cannot put $F_2$ to zero we have to come up with another choice and the simplest one is

$$\mathcal{V} = T_{\mathcal{M}}\,. \tag{2.14}$$

This is called the *standard embedding*, it is automatically stable and the gauge connection is identified with the spin connection. Nevertheless we will see in the next section that the standard embedding is rather constraint for the purpose of model building because the structure group of the tangent bundle is always equal to $SU(3)$ and hence the resulting four-dimensional gauge group will always be $E_6$. In order to construct theories different GUT groups as gauge groups we have to consider more generic vector bundles.

## 2.2 The massless spectrum

In this section we want to see how the massless spectrum of a heterotic Calabi-Yau compactification arises in the supergravity approximation. The ingredients that will be necessary to do so are topological and we can say quite a lot about spectra etc. without the explicite knowledge of the Ricci flat metric that lives on the Calabi-Yau. As we saw in the last section, if we want to obtain $\mathcal{N} = 1$ supersymmetry in four dimensions, we have to compactify on a Calabi-Yau manifold $\mathcal{M}$ coming with a holomorphic and stable vector bundle $\mathcal{V}$

$$\mathcal{V} \to \mathcal{M}\,. \tag{2.15}$$

In the following we will restritct ourselves to the case where in the ten-dimensional theory we have an $E_8 \times E_8$ gauge symmetry rather than an $SO(32)$. Furhtermore one $E_8$ can always be hidden and we will only consider the remaining non-hidden

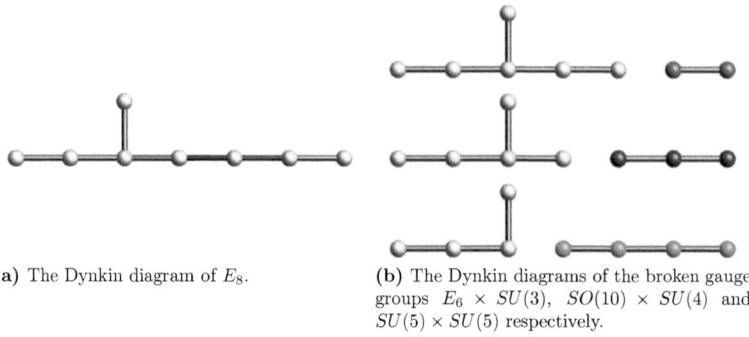

(a) The Dynkin diagram of $E_8$.

(b) The Dynkin diagrams of the broken gauge groups $E_6 \times SU(3)$, $SO(10) \times SU(4)$ and $SU(5) \times SU(5)$ respectively.

**Figure 2.1.:** Breakdown of the gauge group $E_8$ via "breaking" of its Dynkin diagram into an $SU(n)$ part and its commutant.

$E_8$. Compactifying this $10d$ theory on a Calabi-Yau will result in a $4d$ theory in which the visible $E_8$ will be broken by the structure group

$$\mathcal{H} = SU(3),\ SU(4) \text{ or } SU(5) \tag{2.16}$$

of the holomorphic vector bundle $\mathcal{V}$. The unbroken gauge group is then given by the commutant $\mathcal{G}$ of $\mathcal{H}$ in $E_8$ which yields

$$\mathcal{G} = E_6,\ SO(10) \text{ or } SU(5) \tag{2.17}$$

for the gauge group respectively. The case of the standard embedding is here quite a specific one. Since our manifold has $SU(3)$ holonomy, the tangent bundles structure group is forced to be $SU(3)$ and hence the four-dimensional GUT group to be $E_6$. The more general cases of $SO(10)$ and $SU(5)$ GUT groups coming from higher rank vector bundles were first explored by [85–87]. The breakdown of $E_8$ into $\mathcal{G}$ and $\mathcal{H}$ can nicely be read off from the corresponding Dynkin diagram as show in figure 2.1. If we build a physical model this way, eventually we want to calculate its physical quantities. For instance we would like to know what kind of Fermionic fields will appear in the four-dimensional low-energy theory. The relevant Fermionic fields arise then in the decomposition of the gaugiono vector supermultiplet that comes from the breakdown of the $E_8$ gauge group via the structure group $\mathcal{H}$. The relevant representation of $E_8$ here is the **248** which decomposes according to table 2.1. In the large volume limit the only part that will survive the Kaluza-Klein compactification process are the massless modes of the Dirac operator on $\mathcal{M}$. In [85] it was described how to obtain these massless

## 2.2. The massless spectrum

| # zero modes in reps of $H \times G$ | Zero mode are counted by $h^1_{\mathcal{M}}(\bullet)$ of the following bundles: | | | | | |
|---|---|---|---|---|---|---|
| | 1 | $\mathcal{V}$ | $\mathcal{V}^*$ | $\Lambda^2 \mathcal{V}$ | $\Lambda^2 \mathcal{V}^*$ | $\text{End}(\mathcal{V})$ |
| $E_8$ | | | 248 | | | |
| $\downarrow$ | | | $\downarrow$ | | | |
| $SU(3) \times E_6$ | $(\mathbf{1},\mathbf{78}) \oplus (\mathbf{3},\mathbf{27}) \oplus (\overline{\mathbf{3}},\overline{\mathbf{27}})$ | | | | | $\oplus\,(\mathbf{8},\mathbf{1})$ |
| $SU(4) \times SO(10)$ | $(\mathbf{1},\mathbf{45}) \oplus (\mathbf{4},\mathbf{16}) \oplus (\overline{\mathbf{4}},\overline{\mathbf{16}}) \oplus (\mathbf{6},\mathbf{10})$ | | | | | $\oplus\,(\mathbf{15},\mathbf{1})$ |
| $SU(5) \times SU(5)$ | $(\mathbf{1},\mathbf{24}) \oplus (\mathbf{5},\overline{\mathbf{10}}) \oplus (\overline{\mathbf{5}},\mathbf{10}) \oplus (\mathbf{10},\mathbf{5}) \oplus (\overline{\mathbf{10}},\overline{\mathbf{5}})$ | | | | | $\oplus\,(\mathbf{24},\mathbf{1})$ |

**Table 2.1.:** Matter zero modes in representations of the GUT group.

modes simply by calculating certain bundle-valued cohomology groups. The idea is the following: We have to find the massless modes of the Dirac operator. On a complex Kähler manifold we can decompose it with respect to the complex structure as

$$\gamma^i D_i = \gamma^k D_k + \gamma^{\bar{k}} D_{\bar{k}}\,, \tag{2.18}$$

where $D$ is the covariant derivative with respect to the Hermitian metric of the Kähler manifold. One can now show that spinors do in fact correspond to $(0,q)$-forms of the Calabi-Yau manifold, meaning forms that are made of a wedge product of $q$ purely anti-holomorphic one-forms $\sim \mathrm{d}\bar{z}_{\bar{k}}$. Furthermore, the differential d that acts on forms can be decomposed into a holomorphic and anti-holomorphic part

$$\mathrm{d} = \partial + \bar{\partial} \tag{2.19}$$

either as we will explain in a little more detail in chapter 4. Furthermore one can show that acting on spinors with $\gamma^{\bar{k}} D_{\bar{k}}$ corresponds to acting on $(0,q)$-forms with $\bar{\partial}$ and acting with $\gamma^k D_k$ corresponds to acting with $\bar{\partial}^*$ which is the adjoint operator to $\bar{\partial}$ and for the compact case given by

$$\bar{\partial}^* = -*\partial*\,. \tag{2.20}$$

A $(0,q)$-form $\eta$ that represents a zero mode of the Dirac operator must hence obey

$$\bar{\partial}\eta = 0 \quad \text{and} \quad \bar{\partial}^*\eta = 0 \tag{2.21}$$

and is called *harmonic*. If we introduce an equivalence relation in the space of $(0,q)$-forms that identifies all forms that differ only by a form that is given as the derivative of a $(0,q-1)$-form $\eta'$, i.e.

$$\eta \sim \eta + \bar{\partial}\eta' \quad \Rightarrow \quad [\eta] = [\eta + \bar{\partial}\eta']\,, \tag{2.22}$$

one can show that in every equivalence class is exactly one harmonic form. Therefore Counting harmonic forms and hence zero modes of the Dirac operator corresponds to counting equivalence classes of $(0, q)$ forms. The set of all such equivalence classes is called a *cohomology group* in particular the ones we just described, i.e. the "pure" $(0, q)$-forms are taking values in the structure sheaf $\mathcal{O}_\mathcal{M}$ of $\mathcal{M}^1$ and its $q^{\text{th}}$ cohomology group is denoted by

$$H^q(\mathcal{M}; \mathcal{O}_\mathcal{M}) \quad \text{or} \quad H^q_\mathcal{M}(\mathcal{O}_\mathcal{M}). \tag{2.23}$$

Similarly we can also have the $(0, q)$-forms to take values not only in the structure sheaf but rather in some higher rank vector bundle $\mathcal{V}$ and calculate the zero modes of the Dirac operator of these forms which will differ from the former ones. We will call the set of harmonic functions due to the arguments above the $q^{\text{th}}$ vector bundle-valued cohomology groups and denote them by

$$H^q(\mathcal{M}; \mathcal{V}) \quad \text{or} \quad H^q_\mathcal{M}(\mathcal{V}). \tag{2.24}$$

As it turns out the zero modes of the broken four-dimensional gauge group with respect to different representations can then all be found as vector bundle-valued cohomology groups where each representation is determined belongs to a specific bundle. For the chiral spectrum, e.g. in case of bundles with $SU(3)$ structure, the massless modes that correspond to the Fermionic components of chiral superfields of the four-dimensional theory transform in the **27** and $\overline{\mathbf{27}}$ of $E_6$ respectively and their number can be obtained by the dimensions of the vector bundle-valued cohomology groups $\mathcal{V}$ and $\mathcal{V}^*$:

$$n_{\mathbf{27}} = h^1(\mathcal{M}; \mathcal{V}) \quad \text{and} \quad n_{\overline{\mathbf{27}}} = h^1(\mathcal{M}; \mathcal{V}^*). \tag{2.25}$$

Which bundle-valued cohomology contains which zero modes can for all possible cases can be read of from table 2.1. As one can see, there are also gauge singlets that correspond to one-forms with values in endomorphism bundle, i.e. they live in $H^1(\mathcal{M}; \text{End}(\mathcal{V}))$ and are counted by

$$h^1(\mathcal{M}; \text{End}(\mathcal{V})). \tag{2.26}$$

These singlets are basically the infinitesimal deformations of the holomorphic vector bundle and will potentially be moduli of the theory. Besides those there are also the infinitesimal deformations of the Calabi-Yau base manifold which are

---

[1] $\mathcal{O}_\mathcal{M}$ is basically the set of the nowhere vanishing holomorphic functions that play the role of the coefficients of the $(0, q)$-form.

counted by the Hodge numbers, i.e. by the tangent and cotangent bundle-valued cohomology groups:

$$h_{\mathcal{M}}^{2,1} = h^1(\mathcal{M}; T_{\mathcal{M}}) \quad \text{and} \quad h_{\mathcal{M}}^{1,1} = h^1(\mathcal{M}; T_{\mathcal{M}}^*). \tag{2.27}$$

Here the Hodge number $h_{\mathcal{M}}^{2,1}$ counts the possible deformations of the complex structure and $h_{\mathcal{M}}^{1,1}$ counts all possible Kähler deformations. For the standard embedding, (2.25) and (2.27) obviously agree, i.e. the zero modes in the chiral multiplets in the **27** and **$\overline{27}$** correspond to the number of complex structure and Kähler deformations respectively which is really only happening for the standard embedding.

## 2.3 Worldsheet description of the heterotic string & GLSMs

### 2.3.1 The non-linear sigma model:

The description of a perturbative heterotic string compactification is given by a non-linear sigma model (NLSM) that lives in two dimensions and maps the worldsheet to the ten-dimensional target space geometry. The maps

$$X^N : \Sigma \to \mathcal{M} \times \mathbb{R}^{1,3}, \qquad N = 0, ..., 9 \tag{2.28}$$

are the embedding of the worldsheet in the target space $\mathcal{M}$. Furthermore there are Fermions $\psi^N$ and $\gamma^A$ which are right and left-handed worldsheet Fermions respectively. Here the $\Psi^N$ couple to the pullback of the tangent bundle $T_{\mathcal{M}}$ of the target space $\mathcal{M}$ and the $\gamma^A$ couple to the pullback of the vector bundle $\mathcal{V}$ of the target space $\mathcal{M}$. The number of left and right handed supersymmetries can be chosen independently but in order to obtain $\mathcal{N} = 1$ space time supersymmetry we need at least two right handed but no left handed ones. We denote such left-right handed worldsheet supersymmetry by $(0, 2)$. The action of such a sigma model is given in terms of the metric and the $B$-field as

$$S = \frac{1}{4\pi\alpha'} \int d^2z \Big( (G_{MN}(X) + B_{MN}(X)) \partial_z X^M \partial_{\bar{z}} X^N \\ + G_{MN}(X)\psi^M \nabla_z \psi^N + ... \Big) \tag{2.29}$$

Since $G$ and $B$ are rather non-trivial functions in the worldsheet embedding $X$, this action is fairly non-linear and in fact it is quite hard to deal with NLSMs directly and the only way to get the spectrum is in the large volume limit we described above. Furthermore we need our NLSM to be a conformal field theory. Consequently in order to preserve $(0, 2)$ world-sheet super symmetry plus the condition that all diffeomorphism anomalies vanish, namely the model is modular invariant, can the be seen as the following (topological) conditions on the target space

$$c_1(\mathcal{M}) = 0\,, \quad c_1(\mathcal{V}) = 0 \mod 2\,, \quad \operatorname{ch}(\mathcal{M}) = \operatorname{ch}(\mathcal{F}) \qquad (2.30)$$

which are compatible with the conditions derived in the SUGRA picture. For the specific choice of the tangent bundle for $\mathcal{V}$ one can see that now ten of the left-handed Fermions $\gamma^A$ transform in the same way as the right-handed $\Psi^N$ and hence the worldsheet supersymmetry has enhanced to $(2, 2)$.

### 2.3.2 The gauged linear sigma model

Another way to construct super conformal field theories is using a two-dimensional abelian gauge theory, called the *gauged linear sigma model* (GLSM) and was first introduced by Witten in [88]. Let us briefly review a couple of important aspects thereof. For a more thorough introduction we refer to the original literature [88] and extensions [89] or to review articles, e.g. [90,91]. The idea is basically to introduce a two-dimensional abelian gauge theory that is chosen in such a way that it can be traced down to an infra red fixed point, via the renormalization group flow, which has to be conformal. This conformal fixed point will then give us for instance an NLSM that has a Calabi-Yau manifold equipped with a holomorphic vector bundle as target space right in the way we described above. The quantities that do not change under the renormalization group flow can be used to study properties of the NLSM making use of the much easier GLSM. Furthermore the GLSM provides even more general prospects since, depending on the choice of parameters, it will provide scenarios where the target space, i.e. the classical vacuum of the GLSM, cannot be interpreted as a smooth manifold but rather by some singular configuration which might be even only one single point. This means that these parameters give rise to a cone structure which is not unique. The way how to choose this cone structure can then be identified with possible triangulations of a lattice polytope and the different triangulations are called phases of the GLSM [89]. This has a mathematical interpretation as we will see in 3.3 and the triangulations containing the maximal number of cones can be interpreted by smooth geometries. Therefore at low energies, these phases will correspond to NLSMs with a smooth target space geometry and phases corresponding to non-maximal triangulations may correspond to non-geometric Landau-Ginzburg orbifolds or some other more peculiar theories like hybrid models.

## 2.3. Worldsheet description of the heterotic string & GLSMs

More concretely, let us first list the fields in the GLSM equipped with $(0,2)$ supersymmetry, using superspace coordinates $(z, \bar{z}, \theta^+, \theta^-)$. We will have $r$ different $U(1)$ gauge symmetries, labeled by $\alpha = 1, ..., r$. Then there are two sets of chiral superfields:

$$X_i \text{ with } i = 1, \ldots, n \quad \text{and} \quad P_l \text{ with } l = 1, \ldots, n_p. \tag{2.31}$$

In superspace coordinates we can decompose them as

$$X_i = x_i + \theta^+ \psi_i + \theta^+ \bar{\theta}^+ \partial_z x_i, \tag{2.32}$$

$$P_l = p_l + \theta^+ \pi_l + \theta^+ \bar{\theta}^+ \partial_{\bar{z}} p_l. \tag{2.33}$$

Here $x_i$, $p_l$ are the Bosonic components of $X_i$, $P_l$ and the $\psi_{i+}$, $\pi_{i+}$ denote their Fermionic super partners. Furthermore they are charged under the $U(1)^r$ gauge group by

$$Q_i^{(\alpha)} \text{ and } -M_l^{(\alpha)}, \tag{2.34}$$

respectively. Chirality here simply means that the fields obey

$$\bar{D}_+ X_i = 0 = \bar{D}_+ P_l. \tag{2.35}$$

We assume that we can choose $Q_i^{(\alpha)} \geq 0$ as well as $M_l^{(\alpha)}$ and that for each $i$, there exist at least one $r$ such that $Q_i^{(\alpha)} > 0$. Furthermore we have two additional Fermi superfields

$$\Lambda^a \text{ with } a = 1, \ldots, n_\Lambda \quad \text{and} \quad \Gamma_j \text{ with } j = 1, \ldots, c \tag{2.36}$$

that have similar super space expression as (2.32) where the first components are then Fermions $\lambda^a$ and $\gamma_j$ respectively. They are also charged under the $U(1)^r$ gauge symmetry with charges

$$N_a^{(\alpha)} \text{ and } -S_j^{(\alpha)} \tag{2.37}$$

respectively. We also assume that the charges $N_a^{(\alpha)}$ and $S_j^{(\alpha)}$ satisfy the same (semi-)positivity constraints as the $Q_i^{(\alpha)}$ and $M_l^{(\alpha)}$. For the case $(2,2)$ worldsheet supersymmetry, (2.34) and (2.37) agree and the $(0.2)$ chiral and Fermi superfields (2.32) and (2.36) combine to form $(2,2)$ chiral superfields. In the following we will specify such a GLSM by writing all the above data in a table of the form

| $x_i$ | | | | $\Gamma^j$ | | | |
|---|---|---|---|---|---|---|---|
| $Q_1^{(1)}$ | $Q_2^{(1)}$ | $\ldots$ | $Q_n^{(1)}$ | $-S_1^{(1)}$ | $-S_2^{(1)}$ | $\ldots$ | $-S_c^{(1)}$ |
| $Q_1^{(2)}$ | $Q_2^{(2)}$ | $\ldots$ | $Q_n^{(2)}$ | $-S_1^{(2)}$ | $-S_2^{(2)}$ | $\ldots$ | $-S_c^{(2)}$ |
| $\vdots$ | $\vdots$ | $\vdots$ | $\vdots$ | $\vdots$ | $\vdots$ | $\vdots$ | $\vdots$ |
| $Q_1^{(r)}$ | $Q_2^{(r)}$ | $\ldots$ | $Q_n^{(r)}$ | $-S_1^{(r)}$ | $-S_2^{(r)}$ | $\ldots$ | $-S_c^{(r)}$ |

(2.38)

| $\Lambda^a$ | | | | $p_l$ | | | |
|---|---|---|---|---|---|---|---|
| $N_1^{(1)}$ | $N_2^{(1)}$ | $\ldots$ | $N_{n_\Lambda}^{(1)}$ | $-M_1^{(1)}$ | $-M_2^{(1)}$ | $\ldots$ | $-M_{n_p}^{(1)}$ |
| $N_1^{(2)}$ | $N_2^{(2)}$ | $\ldots$ | $N_{n_\Lambda}^{(2)}$ | $-M_1^{(2)}$ | $-M_2^{(2)}$ | $\ldots$ | $-M_{n_p}^{(2)}$ |
| $\vdots$ | $\vdots$ | $\vdots$ | $\vdots$ | $\vdots$ | $\vdots$ | $\vdots$ | $\vdots$ |
| $N_1^{(r)}$ | $N_2^{(r)}$ | $\ldots$ | $N_{n_\Lambda}^{(r)}$ | $-M_1^{(r)}$ | $-M_2^{(r)}$ | $\ldots$ | $-M_{n_p}^{(r)}$ |

(2.39)

The index $\alpha = 1, \ldots, r$ may be suppressed at some points throughout the remainder but will always be there. In the subsequent sections also the notation

$$\mathcal{V}_{N_1,\ldots,N_{n_\Lambda}}[M_1,\ldots,M_{n_p}] \longrightarrow \mathbb{P}_{Q_1,\ldots,Q_n}[S_1,\ldots,S_c] \qquad (2.40)$$

may be used for such a configuration for reasons that will become clear in the following. Anomaly cancellation of the two-dimensional GLSM requires the following set of quadratic and linear constraints to be satisfied

$$\sum_{a=1}^{n_\Lambda} N_a^{(\alpha)} = \sum_{l=1}^{n_p} M_l^{(\alpha)}, \qquad \sum_{i=1}^{n} Q_i^{(\alpha)} = \sum_{j=1}^{c} S_j^{(\alpha)},$$
$$\sum_{l=1}^{n_p} M_l^{(\alpha)} M_l^{(\beta)} - \sum_{a=1}^{n_\Lambda} N_a^{(\alpha)} N_a^{(\beta)} = \sum_{j=1}^{c} S_j^{(\alpha)} S_j^{(\beta)} - \sum_{i=1}^{n} Q_i^{(\alpha)} Q_i^{(\beta)}, \qquad (2.41)$$

for all $\alpha, \beta = 1, \ldots, r$ which corresponds to (2.30).

### 2.3.3 The superpotential

The action of the GLSM contains generically a non-trivial superpotential for the chiral and Fermi super fields which has to be invariant under the $U(1)^r$ gauge

## 2.3. Worldsheet description of the heterotic string & GLSMs

transformations and its most general form is

$$S = \int d^2z d\theta \left[ \sum_j \Gamma^j G_j(X_i) + \sum_{l,a} P_l \Lambda^a F_a{}^l(X_i) \right], \qquad (2.42)$$

where the sub- and superscripts $i$, $j$, $l$, as well as $a$ take values as above and $G_j$ and $F_a{}^l$ are (quasi-)homogeneous polynomials whose multi-degree is fixed by the requirement of gauge invariance of $S$. The multi-degrees of the polynomials $G_j$ and $F_a{}^l$ are given by

$$
\begin{array}{|c|} \hline G_j \\ \hline S_1 \ S_2 \ \ldots \ S_c \\ \hline \end{array}
\quad
\begin{array}{|cccc|} \hline \multicolumn{4}{|c|}{F_a{}^l} \\ \hline M_1 - N_1 & M_1 - N_2 & \ldots & M_1 - N_{n_\Lambda} \\ M_2 - N_1 & M_2 - N_2 & \ldots & M_2 - N_{n_\Lambda} \\ \vdots & \vdots & \vdots & \vdots \\ M_{n_p} - N_1 & M_{n_p} - N_2 & \ldots & M_{n_p} - N_{n_\Lambda} \\ \hline \end{array}.
\qquad (2.43)
$$

For the right-moving Fermions we can derive the following set of Yukawa couplings from the superpotential (2.42):

$$\sum_{i,j} \gamma^j \psi_i \frac{\partial G_j}{\partial x_i}. \qquad (2.44)$$

This way we obtain a mass for a linear combination of the $\psi_i$'s if the derivatives of the hypersurface satisfy a transversality constraint which can be written as

$$\sum_j c_j \frac{\partial G_j}{\partial x_i}(x_1,...,x_n) = 0 \ \forall i \text{ and for some } \vec{c} \neq \vec{0} \qquad (2.45)$$

$$\text{if and only if } x_1 = ... = x_n = 0.$$

If (2.45) was not true we could not guarantee that every linear combination appearing in (2.44) is really present and it would not be clear whether or not the corresponding $\psi$ does obtain a mass indeed. Furthermore the superpotential (2.42) induces couplings for the left-moving Fermions $\lambda^a$ of the Fermi superfields $\Lambda^a$

$$L_{\text{mass}} = \sum_{a,l} \pi_l \lambda^a F_a{}^l \qquad (2.46)$$

which means that a linear combination of the $\lambda^a$ receives a mass by pairing up with one Fermionic component $\pi_l$ of the chiral superfield $P_l$. Again this is only guaranteed if we also put transversality constraints upon the $F_a{}^l$ in the sense that they do not vanish simultaneously

$$\sum_l c_l F_a{}^l(x_1, ..., x_n) = 0 \;\forall a \text{ and for some } \vec{c} \neq \vec{0} \tag{2.47}$$

if and only if $x_1 = ... = x_n = 0$.

We will see below that depending on the choice of the Fayet-Iliopoulos parameters, the set of $x_i$ might be constraint in order to minimize the Bosonic potential which we are about to introduce. In this case the conditions (2.45) and (2.47) might be relaxed a little since the set of $x_i$ may be restricted to form a Calabi-Yau manifold $\mathcal{M}$ given by a constraint in some higher dimensional ambient space. In such a case all the $G_j$ are forced to vanish and define these constraints. Moreover since each $\pi_l$ pairs up with a linear combination of the $\lambda^a$, the massless remaining combinations of the left-moving fermions $\lambda^a$ couple to the kernel of the map $F$ that connects the two spaces of Fermions. Adding the inclusion map we can this way connect three spaces of Fermions in a way such that each map maps Fermions to the kernel of the next one:

$$\mathcal{V} \xrightarrow{\iota} \{\lambda^a\} \xrightarrow{\otimes F_a{}^l} \{\pi_l\}, \tag{2.48}$$

$$\mathcal{V} := \ker F_a{}^l. \tag{2.49}$$

This is not the end of story and once we have a look at the superpotential (2.42) again we can recognize that we are still free to choose an additional gauge symmetry that acts only on the Fermi superfields and leaves the Bosonic chiral superfields unchanged. To define it we introduce $n_F$ new $\Sigma^i$ Fermi superfields that are not chiral, along with the same number of chiral Fermi superfields $\Omega_i$ that are neutrally charged under the $U(1)^r$ gauge group and define the new gauge transformations as

$$\Gamma^j \longrightarrow \Gamma^j + 2E_i^{0j}\Omega^i, \quad \Lambda^a \longrightarrow \Lambda^a + 2E_i^a(X_i)\Omega^i, \quad \Sigma^i \longrightarrow \Sigma^i + \Omega^i, \tag{2.50}$$

for some constants $E_i^{0j}$. This is only true once we impose that also

$$\sum_j E_i^{0j} G_j + \sum_a E_i^a F_a{}^l = 0, \quad \forall i = 1, ..., n_F. \tag{2.51}$$

Furthermore, the $\Sigma^i$ give rise to an extra contribution to the scalar potential which does not play any role for our analysis. The Fermionic components $\sigma^i$ of

## 2.3. Worldsheet description of the heterotic string & GLSMs

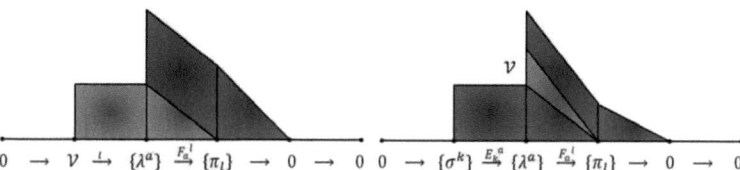

**Figure 2.2.:** Pictorial illustration of the massless Fermions (massless in light gray, massive in dark gray). The horizontal axes display the spaces of Fermions and the vertical one their dimension. On the left hand side the situation of only one kind of mass term is drawn (2.48) and on the right and side the additional Fermionic gaugings are included and $\mathcal{V}$ appears as a quotient space (2.53).

$\Sigma^i$ can now also provide mass terms for the $\lambda^a$ via couplings

$$L_{\text{mass}} = \sum_{i,a} \sigma_i \lambda^a E^i{}_a . \tag{2.52}$$

Hence via this construction the remaining massless left-moving Fermions are those that are in the kernel of $F$ and hence obtain no mass from (2.46), but at the same time do not receive a mass term from (2.52) which means that they do not lie in the image of the map $E$. Hence as we obtain the massless Fermions as a quotient space of $\{\lambda^a\}$:

$$\{\sigma^i\} \xrightarrow{\otimes E^i{}_a} \{\lambda^a\} \xrightarrow{\otimes F_a{}^l} \{\pi_l\} , \tag{2.53}$$

$$\mathcal{V} := \frac{\ker F_a{}^l}{\operatorname{im} E^i{}_a} . \tag{2.54}$$

A pictorial way of seeing the difference between (2.48) and (2.53) is given in figure 2.2.

### 2.3.4 The scalar potential

For the scalar components of the chiral superfields $X_i$ and $P_l$ we have two kind of potentials. An F-term potential and a D-term potential where the latter one will contain as many parameters as $U(1)$ gaugings are present. The F-term scalar potential reads

$$V_F = \sum_{j=1}^{c} \left| G_j(x_i) \right|^2 + \sum_{a=1}^{n_\lambda} \left| \sum_{l=1}^{n_p} p_l \, F_a{}^l(x_i) \right|^2 , \tag{2.55}$$

where $x_i$ and $p_l$ are the Bosonic complex scalars of the corresponding chiral superfields. [2] Introducing the *Fayet-Iliopoulos* parameter $\xi^{(\alpha)} \in \mathbb{R}$ for each $U(1)$

---

[2] Here we would in principle get another scalar contribution due to the additional Fermionic gaugings. But the values of the additional scalar fields coming with such a term are forced to be zero everywhere but at the boundary between phases and since we are only working

the D-term potential can be written as

$$V_D = \sum_{\alpha=1}^{r} \left( \sum_{i=1}^{n} Q_i^{(\alpha)} |x_i|^2 - \sum_{l=1}^{n_p} M_l^{(\alpha)} |p_l|^2 - \xi^{(\alpha)} \right)^2, \qquad (2.56)$$

In order to find the classical vacuum we hence have to minimize these two potentials i.e.

$$V_F = V_D = 0 \qquad (2.57)$$

which in particular implies that

$$\begin{aligned} G_j(x_1,...,x_n) &= 0 \; \forall j \,, \\ \sum_l p_l \, F_a{}^l(x_1,...,x_n) &= 0 \; \forall a \,, \\ \sum_{i=1}^{n} Q_i^{(\alpha)} |x_i|^2 - \sum_{l=1}^{n_p} M_l^{(\alpha)} |p_l|^2 - \xi^{(\alpha)} &= 0 \; \forall \alpha \,, \end{aligned} \qquad (2.58)$$

where the second condition can always rewritten using (2.47) as

$$\{p_1,...,p_{n_p}\} = \{0,...,0\} \quad \text{or} \quad \{x_1,...,x_n\} = \{0,...,0\} \,. \qquad (2.59)$$

### 2.3.5 Geometric interpretation

For a concrete choice of charges one can now determine the classical vacua of the F-term and D-term potential. It turns out that the structure of this vacuum depends crucially on the Fayet-Iliopoulos parameters. In fact the $\mathbb{R}^r$ parametrized by them splits into cones, also called phases, whose boundaries separate different vacuum configurations.

There are certain choices of FI parameters that are particularly interesting and that is for instance where all of them are positive (supposing all the $x_i$ have only positive charges and the $p_l$ have only negative charges). In this case we need some of the $x_i$ not to vanish in order to cancel the $\xi^{(\alpha)}$ and minimize the D-term potential. In order to minimize the F-term, too, the $G_j$ have to vanish and since the $F_a{}^l$ are due to (2.47) not allowed to vanish we need the $p_l$ to be all zero. Hence the D-term vanishing constraints the set of Bosonic fields $x_i$ to obey

$$\mathbb{P}_\Sigma := \left\{ \{x_1,...,x_n\} \; \Big| \; \frac{\sum_{i=1}^{n} Q_i^{(\alpha)} |x_i|^2 = \xi^{(\alpha)}}{U(1)^r} \; \forall \alpha \right\} \qquad (2.60)$$

---

inside proper phases we will simply not write that term explicitly.

## 2.3. Worldsheet description of the heterotic string & GLSMs

and the F-term constraints the set of these $x_i$ further to

$$\mathcal{M} = \{\{x_1, ..., x_n\} \mid \{x_1, ..., x_n\} \in \mathbb{P}_\Sigma \text{ and } G_j(x_1, ..., x_n) = 0 \; \forall j\} \,. \tag{2.61}$$

The space $\mathbb{P}_\Sigma$ is also called a symplectic quotient and the space $\mathcal{M}$ defines a subspace of $\mathbb{P}_\Sigma$. In this case $\mathcal{M}$ will describe a smooth manifold. Furthermore one can show that the Fermions $\sigma$, $\lambda$, $\pi$ do correspond to sections of line bundles that live on that manifold. We will denote them corresponding to their $U(1)$ charges as

$$\mathcal{O}_\mathcal{M}, \; \mathcal{O}_\mathcal{M}(N_a), \; \mathcal{O}_\mathcal{M}(M_l) \,.$$

Furthermore the space of massless Fermions we derived above in (2.48) and (2.53) corresponds to a non-trivial subspace of the direct sum of line bundles that correspond to the $\lambda'_a$s i.e.

$$\mathcal{V} \subset \bigoplus_{a=1}^{n_\Lambda} \mathcal{O}_\mathcal{M}(N_a) \,. \tag{2.62}$$

This will then be a vector bundle[3] and the term non-trivial above means that this vector bundle is no longer a direct sum of line bundles. More precisely, the vector bundle will be given by

$$0 \to \mathcal{V} \xrightarrow{\iota} \bigoplus_{a=1}^{n_\Lambda} \mathcal{O}_\mathcal{M}(N_a) \xrightarrow{\otimes F_a{}^l} \bigoplus_{l=1}^{n_p} \mathcal{O}_\mathcal{M}(M_l) \to 0 \,, \tag{2.63}$$

as $\mathcal{V} = \ker F_a{}^l$ or from

$$0 \to \mathcal{O}_\mathcal{M}^{\oplus n_F} \xrightarrow{\otimes E^i{}_a} \bigoplus_{a=1}^{n_\Lambda} \mathcal{O}_\mathcal{M}(N_a) \xrightarrow{\otimes F_a{}^l} \bigoplus_{l=1}^{n_p} \mathcal{O}_\mathcal{M}(M_l) \to 0 \,, \tag{2.64}$$

as $\mathcal{V} = \frac{\ker F_a{}^l}{\operatorname{im} E^i{}_a}$. Since $\mathcal{V}$ is given by this quotient one can conclude that $\mathcal{V}$ defines a vector bundle of rank

$$\operatorname{rk}(\mathcal{V}) = \#\lambda's - \#\pi's - \#\sigma's = n_\Lambda - n_p - n_F$$

over $\mathcal{M}$.

The construction of a vector bundle this way is well known to mathematicians and referred to as the *monad construction* and we will discuss it in a mathematical framework in 7.4.1.

---

[3]Actually it is also possible the $\mathcal{V}$ is not entirely smooth and hence no longer a vector bundle but a so-called coherent sheaf. On the other hand in all the cases we consider this will not be the case.

**Example 2.3.1** (One single $U(1)$ action). Let us briefly discuss the phases of the GLSM for the most simple choice of a single $U(1)$ and $n_p = 1$. In this case there is only a single Fayet-Iliopoulos parameter and one only obtains two different phases:

For $\xi > 0$ the D-term condition $V_D = 0$ i.e.

$$\sum_{i=1}^{n} Q_i |x_i|^2 - M|p|^2 - \xi = 0 \qquad (2.65)$$

implies that not all $x_i$ are allowed to vanish simultaneously. Thus not all $F_a$ do vanish due to the transversality condition (2.47) and hence vanishing of the F-term potential $V_F = 0$ i.e.

$$\sum_j \left| G_j(x_i) \right|^2 + \sum_a \left| p\, F_a(x_i) \right|^2 = 0 \qquad (2.66)$$

implies $G_j(x_i) = 0$ and $p = 0$. Thus in this phase one gets a $(0,2)$ non-linear sigma-model on a complete intersection of hyper surfaces that belong to the constraining equations $\{G_j = 0\}$ in an ambient space that is given by the symplectic quotient

$$\mathbb{P}_{Q_1,...,Q_n} := \left\{ \frac{\sum_{i=1}^{n} Q_i |x_i|^2 = \xi}{U(1)} \right\}. \qquad (2.67)$$

which is a geometric space where the coordinates are given by the Bosonic fields $x_i$. In fact one can show that this symplectic quotient is equivalent to different construction that is given by a holomorphic quotient. Such holomorphic quotients are well known in mathematics and referred to as *toric varieties*. Many tools to calculate topological quantities of such spaces are known and they are quite convenient to work with. This particular example would correspond to a so-called *weighted projective space* which is a straight forward generalization of a complex projective space. Furthermore the $G_j$ represent (quasi-)homogeneous polynomials that describe the subvarieties

$$\mathcal{S}_j = \{(x_1, ..., x_n) \in \mathbb{P}_{Q_1,...,Q_n} \mid G_j = 0\} \qquad (2.68)$$

which describe, due the transversality constraint (2.45), a complete intersection of hypersurfaces in the ambient space (2.67)

$$\mathbb{P}_{Q_1,...,Q_n}[\mathcal{S}_1, \ldots, \mathcal{S}_c] := \bigcap_{j=1}^{c} \mathcal{S}_j \qquad (2.69)$$

as the target space. In general, the constraints (2.41), guarantee that the complete

## 2.3. Worldsheet description of the heterotic string & GLSMs

intersection defines a threefold with vanishing first Chern class which in our case can be stated by

$$c_1(T_{\mathcal{M}}) = \sum_{j=1}^{c} S_j - \sum_{i=1}^{n} Q_i = 0 \qquad (2.70)$$

and it means that our space $\mathcal{M}$ is a Calabi-Yau manifold. In addition the vector bundle $\mathcal{V}$ is implied to have $SU(n)$ structure group (if it is stable) and the quadratic constraint

$$M^2 - \sum_{a=1}^{n_\Lambda} N_a^2 = \sum_{j=1}^{c} S_j^2 - \sum_{i=1}^{n} Q_i^2 \,, \qquad (2.71)$$

implies the integrated Bianchi-identify

$$c_2(\mathcal{V}) = c_2(T_{\mathcal{M}}) \qquad (2.72)$$

which for more involved cases applies to each geometric phase. Depending on the the number $n_\Lambda$ and on the polynomials $F_a$ we will obtain either the monad

$$0 \to \mathcal{V} \xrightarrow{\iota} \bigoplus_{a=1}^{n_\Lambda} \mathcal{O}_{\mathcal{M}}(N_a) \xrightarrow{\otimes F_a} \mathcal{O}_{\mathcal{M}}(M) \to 0\,, \qquad (2.73)$$

or the monad

$$0 \to \mathcal{O}_{\mathcal{M}}^{\oplus n_F} \xrightarrow{\otimes E^i{}_a} \bigoplus_{a=1}^{n_\Lambda} \mathcal{O}_{\mathcal{M}}(N_a) \xrightarrow{\otimes F_a} \mathcal{O}_{\mathcal{M}}(M) \to 0\,. \qquad (2.74)$$

The vector bundle of our model will therefore be either given as the $\ker(F_a)$ or as $\ker(F_a)/\operatorname{im}(E^i{}_a)$.

The second phase arises for $\xi < 0$. In this case the D-term potential already implies that $\langle p \rangle \neq 0$. Then the F-term potential forces all the $F_a$ to vanish simultaneously which is according to (2.47) equivalent to the vanishing of all the $x_i$. Hence this phase is non-geometric since the classical vacuum is given by a single point obtained from the vev of $p$. Then, the low-energy physics is described by a Landau-Ginzburg (LG) orbifold with a superpotential

$$\mathcal{W}(X_i, \Lambda^a, \Gamma^j) = \sum_j \Gamma^j G_j(X_i) + \sum_a \Lambda^a F_a(X_i). \qquad (2.75)$$

Methods have been developed to deal with such $(0,2)$ LG-models [92, 93], which means in particular the generalization of the BRST methods for the computation of the massless spectrum from $(2,2)$ LG orbifolds to the $(0,2)$ case.

*Remark.* It was first observed in [94] that in this superpotential the constraints $G_j$ and $F_a$ appear on equal footing, so that in particular an exchange of them does not change the Landau-Ginzburg model as long as all anomaly cancellation conditions are satisfied. In [95] this duality was further investigated showing that this exchange is still possible after resolving the generically singular base manifold (see [96] for another kind of $(0,2)$ duality). It is precisely this duality we have studied and we will present our analysis in chapter 9.

In order to be able to analyze GLSM's in their geometric phase it is easier, as already mentioned, to work with the corresponding holomorphic quotients i.e. with toric varieties. A lot is known about them and the next chapters will give techniques to calculate topological quantities thereof. In order to get evidence for a duality it is in particular important to be able to calculate the massless spectrum of the compactification. As argued above the massless fermions will be describing sections in a vector bundle and the way to identify such sections is by calculating cohomology groups. As it turns out at the end we will always be forced to calculate line bundle cohomology which is the reason why we put our effort in developing an algorithm that is able to calculate them fast and a fair part of the following chapters will be devoted to deriving this algorithm as well as the ways to apply it to scenarios we just described, before we turn back to the underlying physics and investigate target space dualities of heterotic compactifications.

# Chapter 3

# Toric Geometry

Toric varieties are a very accessible set of algebraic varieties with the outstanding property that they are completely defined in terms of combinatorial information. In particular also their topological properties are encoded in combinatorics and a lot of these quantities that are usually not calculable for generic algebraic varieties become quite easy to access. Nevertheless, even though all these quantities can be calculated in principle, one can still reach calculable limits once one turns to higher-dimensional complicated cases. In order to push these limits as far as possible it is important to have algorithms of high efficiency in hand. Also vanishing theorems help to avoid unnecessary lengthy calculations. In this chapter we will introduce all the basics on toric geometry required in the remainder of this book.

## 3.1 Physical motivation: Gauged linear sigma models and flux compactifications

Before we start diving deep into the formal mathematics of toric varieties let us first give a little motivation where toric varieties arise in string theory. The most prominent appearance of toric varieties is the one we just explained in chapter 2 in the context of heterotic string theory where they arise as vacuum configuration of the GLSM introduced by Witten in [88]. This two-dimensional field theory is chosen in such a way that there exists an infrared conformal fixed point where it describes the worldsheet of a string, propagating in some geometric or non-geometric background. The vacuum configurations in the geometric case are given by a symplectic quotient which can be shown to coincide with the holomorphic quotient defining toric geometries. Due to that equivalence it is very important to understand the mathematical concepts of toric geometry very well if one wants to work with such theories. Moreover also in the context of type II superstring theories toric varieties can become important once fluxes are introduced in order to stabilize the moduli of the theory. Here complex spaces with an $SU(3)$ structure are required and toric varieties provide a convenient class of such spaces (see e.g. [97] and references thereof).

## 3.2 From projective spaces to toric varieties

In this section we want to introduce the notion of a toric variety in a way one can quickly work with. We will motivate the definition by going through three different examples. We only introduce and use what we actually need later on and therefore the definitions are mathematically not fully rigorous. For a more detailed and very careful introduction to the topic we refer the reader to the book of Cox, Little and Schenck [46].

### 3.2.1 Complex projective spaces

Let us start off with a very simple toric variety namely the complex projective space. Consider $n$ copies of $\mathbb{C}$ as a starting point. The idea of a *complex projective space*, denoted by $\mathbb{CP}^d$ or $\mathbb{P}^d$ is to consider a set of equivalence classes inside $\mathbb{C}^n$ where for the projective spaces we always have $p = n-1$. The equivalence relation identifies all complex straight lines through the origin in this space. For instance consider a point $x \in \mathbb{C}^n$ as an element of $\mathbb{P}^d$, then an arbitrary multiple of this point by a non-zero constant $\lambda \neq 0$ corresponds to one and the same element of $\mathbb{P}^d$. Hence the equivalence relation $\sim$ is defined by

$$(x_1, ..., x_n) \sim (\lambda x_1, ..., \lambda x_n) \; \forall \; \lambda \in \mathbb{C}^* := \mathbb{C}\backslash\{0\} \,. \tag{3.1}$$

So really all points on a straight line are identified by one another with one exception, the origin. Since all these lines intersect at the origin we have to exclude it. Otherwise we would identify all points in our space with the origin and we would never obtain anything non-trivial. For the case $n = 2$ we can actually visualize the space $\mathbb{P}^1$ as shown in figure 3.1a . The action that multiplies every coordinate of a point with a non-zero constant complex number is called a $\mathbb{C}^*$-action for obvious reasons. Since algebraic geometers refer to $\mathbb{C}^*$ as an *algebraic torus* it is also called a torus action. When physicists talk to each other, this action can also take the role of an abelian gauge group acting on some two-dimensional field theory as was explained in chapter 2 and may therefore also be called the $U(1)$ action. Taking everything we just said into account we can quote the proper definition of $\mathbb{P}^d$.

## 3.2. From projective spaces to toric varieties

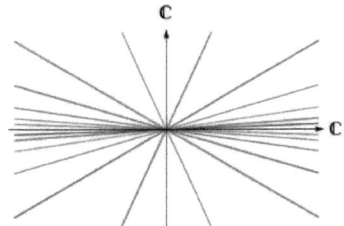
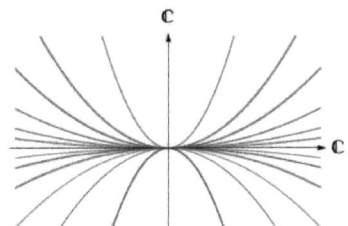

(a) The complex projective space $\mathbb{P}^1$. All the points of the different lines through the origin are identified with one another. The origin itself is excluded.

(b) The weighted projective space $\mathbb{P}_{1,2}$. All points of different parabola through the origin are identified with one another. The origin is excluded.

**Figure 3.1.:** Two examples of a toric variety both having only one $\mathbb{C}^*$-action.

**Definition 3.2.1** (Complex Projective Space). A *complex projective space* $\mathbb{P}^d$ is defined as the quotient space

$$\mathbb{P}^d = \frac{\mathbb{C}^n - \{0\}}{\mathbb{C}^*}, \qquad (3.2)$$

where the $\mathbb{C}^*$-action is defined via the equivalence relation $\sim$:

$$(x_1, ..., x_n) \sim (\lambda^1 x_1, ..., \lambda^1 x_n) \; \forall \; \lambda \in \mathbb{C}^* = \mathbb{C}\backslash\{0\}. \qquad (3.3)$$

A point in $\mathbb{P}^d$ can be written by $(x_1 : ... : x_n)$ (sometimes also simply denoted as $(x_1, ..., x_n)$, keeping (3.3) in mind) where $x_1, ..., x_n$ are then called the *homogeneous coordinates* of $\mathbb{P}^d$.

### 3.2.2 Weighted projective spaces

After defining the complex projective spaces it is quite straight forward to define more general spaces having a $\mathbb{C}^*$-action different from the one of a complex projective space. The picture is more or less the same with just one difference. As we have identified straight lines through the origin to define a complex projective space, the spaces we are considering now are defined by identifying point sets that correspond to, not necessarily linear, polynomials through the origin. This means that we identify points with one another that are related by multiplying every component with the same constant but a different power of this constant, i.e.

$$(x_1, ..., x_n) \sim (\lambda^{Q_1} x_1, ..., \lambda^{Q_n} x_n) \; \forall \; \lambda \in \mathbb{C}^*, \qquad (3.4)$$

where the $Q_i$ are arbitrary numbers in $\mathbb{Z}$ that are called weights. Hence such a space is called a weighted projective space. For different choices of $Q_i$'s we obtain different weighted projective spaces. One simple example for the one-dimensional case is given where $Q_1 = 1$ and $Q_2 = 2$. Then the identification reads for the point $(1,1)$

$$(1,1) \sim (\lambda, \lambda^2) \; \forall \; \lambda \in \mathbb{C}^*. \tag{3.5}$$

This point set is a complex parabola in $\mathbb{C}^2$ and similarly for every starting point different from $(1,1)$ we get a differently shaped parabola which is drawn in figure 3.1b. The formal definition is very much like the one for the complex projective space:

**Definition 3.2.2** (Weighted Projective Space). A *weighted projective space*, denoted by $\mathbb{P}_{Q_1,...,Q_n}$ for $Q_i \in \mathbb{Z} \; \forall \; 1 \leq i \leq n$ is defined as the quotient space

$$\mathbb{P}^d = \frac{\mathbb{C}^n - \{0\}}{\mathbb{C}^*}, \tag{3.6}$$

where the $\mathbb{C}^*$-action is defined via the equivalence relation $\sim$:

$$(x_1, ..., x_n) \sim (\lambda^{Q_1} x_1, ..., \lambda^{Q_n} x_n) \; \forall \; \lambda \in \mathbb{C}^*. \tag{3.7}$$

Now we have already a large number of toric varieties that we can construct using definitions 3.2.1 and 3.2.2. So far we only motivated the construction and visualized their definition. We have not shown any relation to algebraic varieties or smooth manifolds here. In fact one can show that these are algebraic varieties and furthermore that the complex projective spaces are even smooth algebraic varieties and hence they are algebraic manifolds. The weighted projective spaces on the other hand can be shown to be singular spaces and hence are not represented by some smooth manifold. Nevertheless they might contain smooth subvarieties as we will see in section 6. Also, we may perform a blowup of points in the weighted projective space in order to resolve singularities.

### 3.2.3 Toric varieties

The step to a toric variety is not very big now and simply given by an introduction of various $\mathbb{C}^*$-actions inside the same space $\mathbb{C}^n$. An example for a one-dimensional toric variety in $\mathbb{C}^3$ is given by the $\mathbb{C}^*$-actions in the rows of the matrix

$$(Q^\alpha{}_i) = \begin{pmatrix} 1 & 1 & 0 \\ 0 & 2 & 1 \end{pmatrix} \tag{3.8}$$

## 3.2. From projective spaces to toric varieties

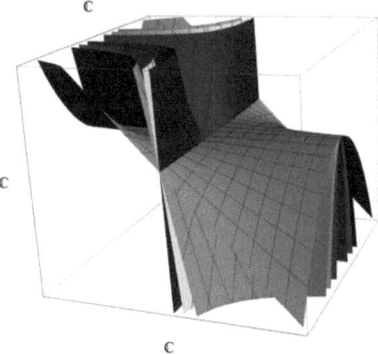

**Figure 3.2.:** The one dimensional toric variety corresponding to the charge matrix in equation (3.8). Since $r = 2$, points on sheets, i.e. complex two-dimensional spaces are identified with one another. Here the exceptional set which has to be removed corresponds to the union of the two horizontal axes.

which gives the following two equivalence relations:

$$(x_1, x_2, x_3) = (\lambda x_1, \lambda x_2, x_3), \quad \forall\, \lambda \in \mathbb{C}^*, \qquad (3.9)$$

$$(x_1, x_2, x_3) = (x_1, \mu^2 x_2, \mu x_3), \quad \forall\, \mu \in \mathbb{C}^*. \qquad (3.10)$$

Plugging in any values for $x_1$, $x_2$ and $x_3$ one can see that these equations just parametrize complex two-dimensional surfaces that span inside $\mathbb{C}^3$ and are built out of two kinds of curves. The curves that correspond to the first row of the matrix (3.8) intersect the horizontal plane at the axis where $x_1$ and $x_2$ equal to zero which is the $x_3$ axis. Similarly intersect the curves that belong to the second row of (3.8) the horizontal plane at the $x_1$ axis. This scenario is plotted in figure 3.2 where the horizontal axes are $x_1$ and $x_3$. Since all the points on these complex surfaces are identified with one another we have to exclude the sets where they intersect in order to get something non-trivial. This set, which is here given by the $x_1$ and the $x_3$ axis is called the *exceptional set*. It is closely related to the so-called *Stanley-Reisner* ideal which will be defined in a second. Let us first quote the general definition of a toric variety which is analog to the one of the projective and weighted projective space:

**Definition 3.2.3** (Toric Variety). Let $d$, $r \in \mathbb{N}$ and $n = d + r$. A $d$ dimensional *toric variety* $\mathbb{P}_\Sigma$ is defined as the quotient space

$$\mathbb{P}_\Sigma = \frac{\mathbb{C}^n - Z}{(\mathbb{C}^*)^r}, \tag{3.11}$$

where $Z$ is the exceptional set. The $(\mathbb{C}^*)^r$ actions correspond to the equivalence relations that are given by a matrix $(Q^\alpha{}_i)$ which is defined according to

$$(x_1, ..., x_n) \sim (\lambda^{Q^\alpha{}_1} x_1, ..., \lambda^{Q^\alpha{}_n} x_n) \ \forall \ \alpha = 1, ..., r, \ \forall \ \lambda \in \mathbb{C}^*, \tag{3.12}$$

which identifies points on $r$-dimensional subspaces in $\mathbb{C}^n$.

*Remark.* The exceptional set is a crucial part of the toric variety that encodes a lot of information on its topology. It basically tells us which combination of homogeneous coordinates are not allowed to vanish simultaneously. In the framework of the GLSM this task corresponds to minimizing the Bosonic potential which was done by solving equations (2.58). The exceptional set is here given by the solution of equation three in (2.58) and this relation between toric varieties and the classical vacuum of a two-dimensional field theory helps many physicists to loose their fear of these spaces.

What we specifically usually need is the quite closely related notion of the following.

**Definition 3.2.4** (Stanley-Reisner Ideal). Let $Z$ be the exceptional set of a toric variety $\mathbb{P}_\Sigma$. The *Stanley-Reisner ideal* $I_\Sigma$ is the minimal ideal containing square-free monomials corresponding to the different subsets of the exceptional set:

$$I_\Sigma := \langle \ \mathbf{x}^\tau \ | \ \{\mathbf{x}_\tau = 0\} \subset Z \rangle, \tag{3.13}$$

where we defined $\tau := \{i_1, ..., i_k\}$ along with

$$\mathbf{x}^\tau := \prod_{j=1}^{k} x_{i_j} \quad \text{and} \quad \mathbf{x}_\tau := \{x_{i_1}, ..., x_{i_k}\} \tag{3.14}$$

**Example 3.2.1.** For the complex projective space and the weighted projective space we had only to remove the origin. So the exceptional set $Z$ has only this one defining subset:

$$Z = \{(x_1, x_2, ..., x_n) \in \mathbb{C}^n \ | \ (x_1, x_2, ..., x_n) = (0, 0, ..., 0)\}. \tag{3.15}$$

Therefore we get a Stanley-Reisner Ideal which contains only one monomial:

$$I_\Sigma = \langle x_1 \cdot x_2 \cdot \ldots \cdot x_n \rangle. \tag{3.16}$$

**Example 3.2.2.** As a second example let us consider the toric variety defined by the matrix in (3.8) above. As we saw, we had to remove the $x_1$ and the $x_3$ axes. Therefore the exceptional set of this example is

$$Z = \{(x_1, x_2, x_3) \in \mathbb{C}^3 : (x_2, x_3) = (0,0) \text{ or } (x_1, x_2) = (0,0)\}, \tag{3.17}$$

and from that we obtain the Stanley-Reisner ideal which has two elements:

$$I_\Sigma = \langle x_2 \cdot x_3, \ x_1 \cdot x_2 \rangle. \tag{3.18}$$

As one can imagine, these sets might get very complicated to derive once we choose a matrix $Q$ that does not look as nice as the one in (3.8) and at some point it is impossible to simply read off the exceptional set and hence the Stanley-Reisner ideal. But there is a very convenient way to calculate these sets systematically in terms of $d$ dimensional polytopes which we are going to explain in the next section.

*Remark.* We have not said anything about smoothness of a generic toric variety so far. While we know that the projective spaces are always smooth and the weighted projective spaces are always singular, for a generic toric variety it is a priory not clear. There are smooth ones that are in particular no products of complex projective spaces and there are also singular ones which are not products of weighted projective spaces. How one can check for smoothness of a toric variety in a combinatorial way will also be content of the next section.

## 3.3 Lattice polytopes and fans

After we learned in the last section what the idea of a toric variety is and already saw a few fairly simple examples, we want now to turn to methods and techniques that allow us to handle more complicated toric varieties. In fact we want to have an explicit description of a method to calculate the Stanley-Reisner ideal and hence also the exceptional set for in principle arbitrarily complicated toric varieties.

## 3.3.1 Toric varieties from lattice polytopes

In order to achieve that we will introduce a very useful description of the combinatorial data of the toric variety as an $n$ dimensional lattice polytope. The identification is as follows: Consider a $d$ dimensional toric variety given by definition 3.2.3. Now consider a $d$ dimensional lattice

$$M := \mathbb{Z}^d \qquad (3.19)$$

and define a set of vectors in this lattice as follows:

$$\left\{ w \in M : \sum_{i=1}^{n} Q^{\alpha}{}_i w_i = 0 \right\}. \qquad (3.20)$$

This is nothing else but the kernel of the matrix $Q$ that defines our toric variety. This set of vectors in the lattice $M$ can be interpreted as an $d$ dimensional lattice polytope as follows: The set of vertices denoted by $\Delta^\circ$ is given by the rows of the transposed kernel of the matrix $Q$

$$\Delta^\circ := \ker(Q). \qquad (3.21)$$

Since $\Delta^\circ$ is just the kernel of the $Q$-matrix, the dimension of the polytope can always be identified with the complex dimension of the toric variety. In fact in order to obtain information about the topology we usually need both the polytope $\Delta^\circ$ and the matrix $Q$. In the literature usually the polytope is the data provided and from that we can also obtain the matrix $Q$ as the kernel of $\Delta^\circ$:

$$Q = \ker(\Delta^{\circ \mathrm{T}})^{\mathrm{T}} \qquad (3.22)$$

and in fact to define it this way round is a bit more general since it may happen, e.g. for non-compact toric varieties, that there is a polytope defining the toric variety in a slightly different way as we did above that does not give rise to a matrix $Q$. But in all cases we consider this will always be possible.

**Example 3.3.1.** Let us consider as an easy example the two-dimensional space $\mathbb{P}^2$ living as a quotient in $\mathbb{C}^3$ hence having one equivalence relation. It is is given by the quotient (3.2.3) where $Q$ is given by

$$Q = (1, 1, 1). \qquad (3.23)$$

## 3.3. Lattice polytopes and fans

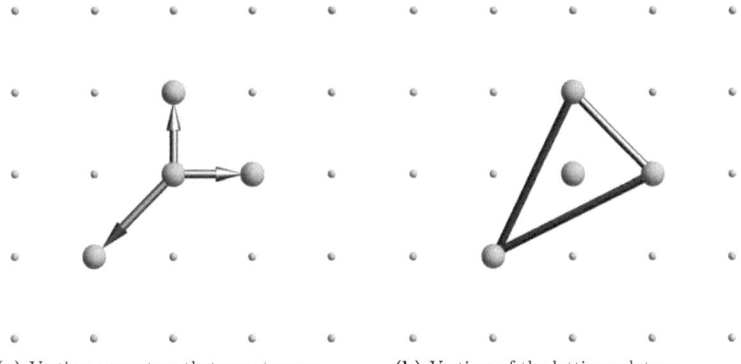

(a) Vertices as vectors that sum to zero    (b) Vertices of the lattice polytope

**Figure 3.3.:** The toric diagram of $\mathbb{P}^2$; the vertex in the center of the image marks the origin $(0,0)$. The sum of all three vectors with coefficients 1 is zero. Hence we have the 1-weighted projective space $\mathbb{P}_{1,1,1} \equiv \mathbb{P}^2$.

Therefore we find the transposed kernel to be

$$\Delta^\circ = \ker\{(1,1,1)\} = \begin{pmatrix} -1 & -1 \\ 0 & 1 \\ 1 & 0 \end{pmatrix}. \tag{3.24}$$

Hence we got three vertices for the corresponding polytope,

$$v_1 = (1,0), \qquad v_2 = (0,1), \qquad v_3 = (-1,-1). \tag{3.25}$$

They are plotted in figure 3.3a. In quite the same way one could construct the polytope of a weighted projective space e. g. $\mathbb{P}_{1,1,2}$ by simply moving the vertex $(-1,-1)$ to $(-1,-2)$. This scenario is shown in figure 3.4 and as one can observe that there is now one point, namely the point $(0,-1)$ that actually lies inside the polytope without being a vertex of it. We will see later on that precisely that is what causes the toric variety to be singular.

### 3.3.2 Cones and fans

So far we only considered examples where the toric variety is uniquely described by the matrix $Q$ or equivalently the polytope $\Delta^\circ$. It turns out that this is actually not quite enough in general. As we defined the toric variety in (3.2.3), we explained how the equivalence relation $\sim$ can be defined in terms of the matrix

# 3. Toric Geometry

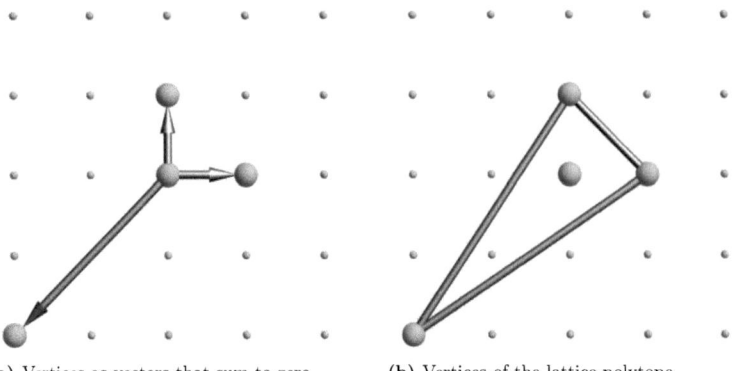

(a) Vertices as vectors that sum to zero

(b) Vertices of the lattice polytope

**Figure 3.4.:** The toric diagram of $\mathbb{P}_{122}$; the vertex in the center of the image marks the origin $(0,0)$. The sum of all three vectors with coefficients 1, 2 and 2 respectively is zero. The lattice point in 3.3b that lies inside the polytope corresponds to a singularity of the space.

Q. At that time we always supposed that we know how the exceptional set looks like and in the simple examples we considered, it was always possible to just write it down uniquely. In fact this choice is in general not at all unique and one might have even hundreds equally valid ways to choose this set. Every such choice will then stand for an a priory topologically distinct toric variety. In this subsection we want to see how we can systematically derive all possible choices of this exceptional set and also how we can compute certain topological properties from it.

**Definition 3.3.1** (Cone). We define a *(strongly convex rational polyhedral) cone* $\sigma$ to be a set spanned by a finite number of vectors $v_1, ..., v_k$ like

$$\sigma = \left\{ \sum_{i=1}^{k} a_i v_i : a_i \geq 0 \right\}, \qquad (3.26)$$

with the property that $\sigma \cap (-\sigma) = 0$. Hence the apex of each such cone is always the origin. A cone is called *simplicial* if its dimension equals the number of generating vectors, i.e. $k = \dim(\sigma)$.

**Definition 3.3.2** (Fan). A *fan* $\Sigma$ is a collection of cones such that for every two cones $\sigma_i, \sigma_j \in \Sigma$, the intersection $\sigma_i \cap \sigma_j$ is also a cone which is itself an element of $\Sigma$. By $\Sigma(k)$ we denote the subset of all $k$-dimensional cones in $\Sigma$.

We have seen in the last subsection that we can define a polytope representing information of the toric variety by taking the kernel of the Q-matrix. Therefore

## 3.3. Lattice polytopes and fans

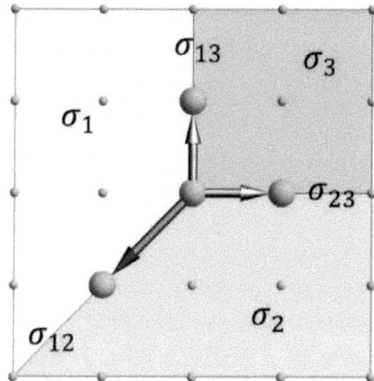

**Figure 3.5.:** The fan of $\mathbb{P}^2$. There are three two-dimensional cones $\sigma_i$, three one-dimensional cones $\sigma_{ij}$ and one zero-dimensional cone, the origin.

every vertex of the polytope is associated to one homogeneous coordinate of the variety. In figure 3.5 we can see, exemplary for $\mathbb{P}^2$, that such a polytope gives rise to a fan where the origin is the apex, the vectors to the vertices span the one-dimensional cones and combinations of two vectors span the maximal cones which are in this case the three two-dimensional cones. The idea is now that the different regions of the fan tell us where on the toric variety we are located in terms of which coordinates may or may not vanish in that particular region. Every maximal cone corresponds then to an affine toric variety which is simply a topologically trivial one and can be described by one single patch $\mathcal{U}_i$ i.e. by one single copy of $\mathbb{C}^2$ in case of $\mathbb{P}^2$. In practice these are the regions where some of the homogeneous coordinates of the variety are allowed to take any value and the remaining ones are not allowed to vanish according to the following rule:

A coordinate $x_i$ is **not** allowed to vanish on the subset $\mathbb{P}_\Sigma|_{U_\sigma}$ of the toric variety belonging to the cone $\sigma \in \Sigma$, if the one-dimensional cone $v_i$, corresponding to $x_i$, is not a generator of $\sigma$.

A fan therefore provides instructions on how these affine toric varieties are to be glued together consistently. Here the lower-dimensional cones encode along which loci the affine toric varieties are glued together and the resulting topology may then no longer be trivial, i.e. the resulting toric variety may no longer be an affine one. Since the information of which homogeneous coordinates are allowed to vanish on the loci corresponding to the cones is completely determined by this construction, the problem of finding the exceptional set is just the problem of finding a fan for a given polytope. Those vectors that do not span any cone of the fan correspond to coordinates that are never allowed to vanish simultaneously on $\mathbb{P}_\Sigma$:

$$\{x \in \mathbb{C}^N : (x_{i_1},...,x_{i_k}) = (0,...,0)\} \subset Z$$
$$\Leftrightarrow \{v_{i_1},...,v_{i_k} \in \Sigma(1)\} \text{ do not span a cone in } \Sigma, \quad (3.27)$$

were $v_{i_k}$ is the one dimensional cone associated to the homogeneous coordinate $x_{i_k}$ of $\mathbb{P}_\Sigma$.

**Example 3.3.2.** Let us come back to our example of $\mathbb{P}^2$ to see how it explicitly works. As shown in figure 3.5 we have three cones of dimension two, three cones of dimension one and the origin as zero dimensional cone. The open cover of $\mathbb{P}^2$ corresponding to the maximal cones is then given by

$$\mathcal{U}_1 := \{(x_1,x_2,x_3) \in \mathbb{P}^2 \mid x_1 \neq 0\}, \text{ with local coords. } \left(\frac{x_2}{x_1},\frac{x_3}{x_1}\right) \in \mathbb{C}^2,$$

$$\mathcal{U}_2 := \{(x_1,x_2,x_3) \in \mathbb{P}^2 \mid x_2 \neq 0\}, \text{ with local coords. } \left(\frac{x_1}{x_2},\frac{x_3}{x_2}\right) \in \mathbb{C}^2,$$

$$\mathcal{U}_3 := \{(x_1,x_2,x_3) \in \mathbb{P}^2 \mid x_3 \neq 0\}, \text{ with local coords. } \left(\frac{x_1}{x_3},\frac{x_2}{x_3}\right) \in \mathbb{C}^2,$$

where $\mathcal{U}_i$ corresponds to the maximal cone $\sigma_i$. According to the fan given in figure 3.5, these patches are glued together along loci

$$\mathcal{U}_{ij} := \mathcal{U}_i \cap \mathcal{U}_j = \{(x_1,x_2,x_3) \in \mathbb{P}^2 \mid x_i \neq 0, \ x_j \neq 0\} \quad (3.28)$$

which correspond to the one-dimensional cones $\sigma_{ij}$ and furthermore connected via the one zero-dimensional cone $\sigma_{123}$ that corresponds to the intersection of all three patches

$$\mathcal{U}_{123} := \mathcal{U}_1 \cap \mathcal{U}_2 \cap \mathcal{U}_3 = \{(x_1,x_2,x_3) \in \mathbb{P}^2 \mid x_1 \neq 0, \ x_2 \neq 0, \ x_2 \neq 0\}. \quad (3.29)$$

So we can see that we always find patches where any combination of homogeneous coordinates are allowed to vanish simultaneously with the only exception of the case where they all vanish at at the same time.

**Definition 3.3.3** (Stanley-Reisner ideal revisited). Using these notions we can define the *Stanley-Reisner ideal* as

$$I_\Sigma = \left\langle \prod_{j=1}^m x_{i_j} \mid \langle v_{i_1},...,v_{i_m}\rangle \notin \Sigma \right\rangle. \quad (3.30)$$

## 3.4. Line bundles and divisors

We furthermore define the *irrelevant ideal* which we will need later on as

$$B_\Sigma = \left\langle \prod_{\substack{j \neq i_k, \\ k=1,...,m}}^{m} x_j \mid \langle v_{i_1},...,v_{i_m} \rangle \in \Sigma \right\rangle, \qquad (3.31)$$

to be generated by monomials that correspond to complements of cones.

**Definition 3.3.4** (Triangulation). Let $\Delta^\circ$ be an $d$-dimensional lattice polytope. A *triangulation* of $\Delta^\circ$ is a partition of this polytope into maximal-dimensional cones that form a fan. A *maximal triangulation* $\mathcal{T}$ of $\Delta^\circ$ is a triangulation such that no other triangulation $\mathcal{T}'$ of $\Delta^\circ$ has more maximal cones than $\mathcal{T}$.

Each maximal triangulation $\mathcal{T}$ of a polytope $\Delta^\circ$ gives therefore rise to a toric variety and since the maximal cones are glued together differently in two different maximal triangulations of the same polytope, they will give rise to different topological spaces. Tools to calculate the triangulations that belong to a polytope as well as its Stanley-Reisner ideal are for instance [58], [59] and a nice interface to both [64]. For model building in string-theory we will be faced with toric varieties of at least three dimensions and the techniques even though straightforward get very tedious and it is in fact not possible to avoid tools like these.

## 3.4 Line bundles and divisors

In this last section of the toric geometry chapter we want to introduce the notion of coherent sheaves as well as line bundles, divisors and their relation to each other. All these objects are closely related to the topology of the ambient toric variety and can be used as tools to calculate such topological quantities. How this can be done explicitly will be dedicated to later chapters.

**Definition 3.4.1** (Holomorphic line bundles and sheaves). A rank one vector bundle $\mathcal{L}$ over a toric variety $\mathbb{P}_\Sigma$ that has holomorphic transition functions is called a *holomorphic line bundle*[1]. A *section* or *global section* of a line bundle is defined to be a map that maps every base point to a fiber of the line bundle which is isomorphic to $\mathbb{C}$:

$$s : \mathbb{P}_\Sigma \longrightarrow \mathcal{L}, \qquad (3.32)$$

$$s(x) \in \mathcal{L}_x \cong \mathbb{C}. \qquad (3.33)$$

---
[1] Throughout the remainder we might drop the adjective holomorphic from time to time, but those line bundles are always understood to be holomorphic.

This means every point $x \in \mathbb{P}_\Sigma$ takes a value in some complex line over the base. Therefore such a section can be represented by a polynomial in the homogeneous coordinates:

$$s(x) = \sum_{a_1,\ldots,a_n} A_{a_1,\ldots,a_n} x_1^{a_1} \cdot \ldots \cdot x_n^{a_n}, \tag{3.34}$$

with some coefficients $A_{a_1,\ldots,a_N} \in \mathbb{C}$. Since the homogeneous coordinates are only defined up to $r$ different $U(1)$-actions given by $Q$, we can rewrite (3.34) as

$$s(x) = \sum_{a_1,\ldots,a_n} A_{a_1,\ldots,a_n} \lambda^{a_1 \cdot Q_1^\alpha} \cdots \lambda^{a_n \cdot Q_n^\alpha} x_1^{a_1} \cdot \ldots \cdot x_n^{a_n 1}, \quad \forall \lambda \in \mathbb{C}. \tag{3.35}$$

So if we want this map to be a well defined section in a rank one bundle, we cannot allow arbitrary terms in (3.35) but we have to restrict to those such that the sum of the $U(1)$ charges of the coordinates always gives a unique number $Q_{\text{deg}}^\alpha$ for every $\alpha$, i.e. the following coefficients have to vanish:

$$A_{a_1,\ldots,a_n} = 0 \quad \text{if} \quad \lambda^{a_1 \cdot Q_1^\alpha + \cdots + a_n \cdot Q_n^\alpha} \neq \lambda^{Q_{\text{deg}}^\alpha} \text{ for some } \alpha. \tag{3.36}$$

A polynomial that satisfies (3.36) for all $r$ $U(1)$-charges $Q_{\text{deg}}^\alpha$ is called *homogeneous* and the $r$ numbers $Q_{\text{deg}}^\alpha$ are called the *multi degree* of the homogeneous polynomial.

In contrast to a vector bundle that attaches a vector space to each base point, there is the related concept of a so-called *sheaf* that attaches a group to each open set of the base space instead. In fact for the precise definition one has of course to put some restrictions on this to make sure that it behaves fine by intersecting open sets etc. but we will not state these formal definitions here since we will not need them. The important point for us is that the set of sections in a vector bundle that are defined on the open cover of $\mathbb{P}_\Sigma$ define a sheaf. Therefore in the remainder if we talk about sheaves we always mean the sheaf of sections associated to the corresponding vector bundle.

*Remark.* The set of homogeneous polynomials form a vector space and in chapter 4 we will see that the dimension of this vector space is a topological quantity referred to as cohomology of the line bundle. Here we will see that besides the homogeneous polynomials, also so-called homogeneous Laurent polynomials will become important. These differ from the homogeneous polynomials only by the fact that the terms are no longer monomials but quotients of monomials. So the homogeneous multi degree may besides positive numbers and zero also contain negative numbers. So there may also appear cases where no global sections can be written down at all and only Laurent polynomials will determine the topology

## 3.4. Line bundles and divisors

of the line bundle.

**Definition 3.4.2.** As we saw in 3.4.1, the sections of a line bundle have to have a certain multi degree $Q^\alpha_{\text{deg}}$. This multi degree is also called the *first Chern class* and one can show that in smooth cases it identifies line bundles that are isomorphic to each other. Hence from now on we will often simply refer to a line bundle $\mathcal{L}$ over $\mathbb{P}_\Sigma$ with first Chern class $Q^\alpha_{\text{deg}}$ as

$$\mathcal{L} =: \mathcal{O}_{\mathbb{P}_\Sigma}(Q^\alpha_{\text{deg}}) \,.$$

**Definition 3.4.3** (Toric and effective divisors, Picard group). Since we defined line bundles over toric varieties we would now like to define the closely related concept of divisors in $\mathbb{P}_\Sigma$. A *divisor*[2] on a toric variety defined as the formal sum of irreducible codimension one subvarieties, i.e.

$$\mathcal{D} = \sum_{i=1}^{n} a_i \mathcal{D}_i \,, \ a_i \in \mathbb{Z}\,, \text{ where} \tag{3.37}$$

$$\mathcal{D}_i := \{x \in \mathbb{P}_\Sigma \mid x_i = 0\} \,. \tag{3.38}$$

The subvarieties $\mathcal{D}_i$ are called *toric divisor*. Since a divisor is a sum of codimension one subvarieties it should be possible to obtain it (maybe patch wise) by one single equation in the homogenous coordinates that is put to zero. This seems similar than defining the sections of a line bundle where we also needed $\mathbb{C}$-valued functions to define them. It turns out that there is actually precise correspondence of line bundles and divisors and we can use either language to describe these objects. The homogeneous multi degree of the equation that we use to define the divisor is then of course given by the first Chern class of the corresponding line bundle.

Due to our linear relations between the coordinates we can see that equation that defines the divisor in (3.37) is not always unique and that we might get a linear equivalence between some of them. As one might guess it turns out that the number of independent divisors, also called the *Picard group*, is just given by the number of $U(1)$ charges defining the toric ambient variety. Therefore the divisor will only depend on $r$ different constants defining the formal sum of subvarieties. If all these constants are larger or equal to zero, we call the divisor *effective*.

---

[2] In the remainder we will mostly be dealing with smooth manifolds and hence we will not use the distinction between so- called Weyl and Cartier divisors since they are the same is such a setting.

# Chapter 4

## Cohomology of Line Bundles

In the last section we have introduced divisors and line bundles and argued why they are basically the same. In this chapter we want to present one of the main results we accomplished. Since line bundles and divisors naturally appear at various places in string theory it is quite important to better understand the topology of these objects. For this purpose we have to calculate their cohomology. This task is in principle possible but it can take a lot of time doing by hand. Furthermore the so far known ways to compute them via plain Čech complexes take into account all open sets of the corresponding base space and can, hence, become very lengthy. Especially for our purposes that involve calculating a large number of them one needs more efficient ways as well as vanishing theorems to keep the time efforts manageable. That is exactly the goal of this chapter.

## 4.1 Physical motivation: Heterotic GUTs, line bundles on D7 branes and type IIB/F-theory instantons

Line bundles can be used to calculate many topological quantities of toric varieties. They can also be used to obtain the topological structure of vector bundles over such spaces or subspaces and are therefore very useful for heterotic string model building as we saw in chapter 2. For chapter 9, scans of large sets of geometries were performed and needed an efficient way to calculate the basic ingredient of these topologies, i.e. the line bundle cohomology. But also in type II string theory line bundles find their application since they are carried by $D7$-branes. The cohomology of these line bundles is crucial to the construction, see for instance the review [28] for more details. Furthermore in the type IIB setting and also its non-perturbative version, F-theory, instantons are present and manifest themselves as zero modes of certain line bundles that live on three- and four-folds respectively [98, 99].

## 4.2 Cohomology: The idea

Let us first of all talk about the idea behind the word *cohomology* and try to develop an intuition for what it stands.

### 4.2.1 De Rham cohomology

In the probably most intuitive case, the topology we would like to investigate is determined by the cotangent bundle $T^*_\mathcal{M}$ and hence determines the "shape" i.e. the topology of a base manifold $\mathcal{M}$ itself. Here calculating cohomology groups means that we consider differential forms of a certain degree $p$ which are nothing but sections in the $p^{\text{th}}$ exterior power of the cotangent bundle

$$\bigwedge_{i=1}^{p} T^*_\mathcal{M} =: \Omega^p(\mathcal{M}). \tag{4.1}$$

Then we look at those forms that are closed with respect to the differential

$$\mathrm{d} := \sum_k \mathrm{d}x_k \wedge \frac{\partial}{\partial x_k} \tag{4.2}$$

but not exact, i.e. are not the derivative of another form. For a $p$-form $\omega_p$ for instance this would mean that

$$\mathrm{d}\omega_p = 0 \quad \text{and} \quad \nexists\ (p-1)\text{-form } \omega_{p-1} \text{ s.t. } \mathrm{d}\omega_{p-1} = \omega_p. \tag{4.3}$$

This set of forms is called the $p^{\text{th}}$ *de Rham cohomology group* of $\mathcal{M}$ and we will denote it by $H^p_{\mathrm{dR}}(\mathcal{M})$.

How does this set of differential forms tell us anything about the topology of the bundle? In order to get an idea why this is indeed the case, we consider an example.

**Example 4.2.1.** Let $\mathcal{M}$ be the two-torus $T^2$ and $(\theta, \phi)$ be local coordinates that parametrize the two circles, $\theta, \phi \in [0, ..., 2\pi)$. In order to specify its topology let us derive its cohomology. Since its real dimension is two, we can at most write down an anti symmetric tensor with two indices. Hence we have at most two-forms on this space. Let us start with $H^0_{\mathrm{dR}}(T^2)$. We are looking for smooth functions $\omega_0$ with vanishing differential. Since there are no differential forms with negative degree, every closed function is non-exact, therefore lie in the cohomology and

## 4.2. Cohomology: The idea

must satisfy

$$d\omega_0 = \frac{\partial \omega_0}{\partial \theta} d\theta + \frac{\partial \omega_0}{\partial \phi} d\phi = 0. \tag{4.4}$$

This can only be the case if $\omega_0$ is a constant in $\mathbb{R}$ and hence the dimension of the $0^{\text{th}}$ de Rham cohomology is simply the number of connected components of the manifold which is in our case one:

$$\dim_{\mathbb{R}}(H^0_{\text{dR}}(T^2)) =: h^0(T^2) = 1. \tag{4.5}$$

For $H^1_{\text{dR}}(T^2)$ we have to find all closed non-exact one-forms. The most general one-form is given by

$$\omega_1 = f \, d\theta + g \, d\phi \in \Omega^1(T^2) \quad f, g \in C^\infty(T^2). \tag{4.6}$$

The requirement of closeness of $\omega_1$

$$d\omega_1 = \frac{\partial f}{\partial \phi} d\phi \wedge d\theta + \frac{\partial g}{\partial \theta} d\theta \wedge d\phi = 0 \tag{4.7}$$

forces the coefficient functions to be either constants or to satisfy

$$\frac{\partial f}{\partial \theta} = \frac{\partial g}{\partial \phi}. \tag{4.8}$$

In fact one can show that the second possibility represents either an exact one-form, or it differs from the case where $f = f_0 \in \mathbb{R}$ and $g = g_0 \in \mathbb{R}$ by an overall constant. Furthermore $d\theta$ and $d\phi$ are non-exact, because $\theta$ and $\phi$ are multi-valued due to the fact that they parameterize the two circles of the torus. Hence we can write an arbitrary one-form in $H^1_{\text{dR}}(T^2)$ as

$$\omega_1 = f_0 \, d\theta + g_0 \, d\phi \in H^1_{\text{dR}}(T^2) \quad (f_0, g_0) \in \mathbb{R}^2 \tag{4.9}$$

which spans a vectorspace with real dimension $h^1_{\text{dR}}(T^2) = 2$. In a similar way we can find that the dimension of $H^2_{\text{dR}}(T^2)$ is again one. The dimensions of the de Rham cohomology groups are also referred to as *Betti numbers*.

This example already demonstrates that the cohomology groups that can be counted by sections into certain bundles give information about the topology and hence the shape of the underlying space. Let us now define the de Rham cohomology in a little more rigorous way. To do that we need a couple of other definitions

**Definition 4.2.1** (Complex, exact sequence, cohomology). We define a *complex*

# 4. Cohomology of Line Bundles

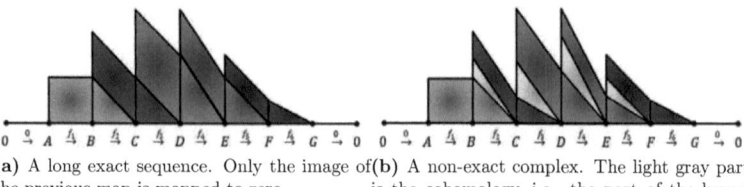

(a) A long exact sequence. Only the image of the previous map is mapped to zero

(b) A non-exact complex. The light gray part is the cohomology, i.e. the part of the kernel that lies not in the image of the previous map.

**Figure 4.1.:** Pictorial representation of a long exact sequence and a long complex with cohomology. On the horizontal line the spaces are shown whereas the vertical direction indicates the dimension of the space. The connecting between two spaces in the sequence indicates the image of the map. The fact that $f_i \circ f_{i+1} = 0$ can be seen nicely.

to be a sequence of vector bundles $A, B, C, D...$, connected via maps $f_1, f_2, f_3, f_4...$,

$$\cdots \longrightarrow A \xrightarrow{f_1} B \xrightarrow{f_2} C \xrightarrow{f_3} D \xrightarrow{f_4} \cdots \qquad (4.10)$$

in a way such that

$$\text{im}(f_i) \subset \ker(f_{i+1}), \forall i \quad \Leftrightarrow \quad f_i \circ f_{i+1} = 0. \qquad (4.11)$$

If the image of one map is always precisely equal to the kernel of the following map, i.e.

$$\text{im}(f_i) = \ker(f_{i+1}) \forall i, \qquad (4.12)$$

we say the complex is *exact*. A pictorial way to imagine complexes can be found in figure 4.1. For non-exact complexes, we define the failure of the complex to be exact as the *cohomology* of the complex at that specific position. For instance the cohomology of the complex (4.10) at position $B$ is given by

$$\frac{\ker(f_2)}{\text{im}(f_1)}. \qquad (4.13)$$

As it turns out everything that we ever call cohomology is cohomology in this sense. This means that we will always have some complex that contains the cohomology as the measure of the failure of this complex to be an exact sequence.

**Definition 4.2.2** (De Rham Complex). As an example we take the de Rham cohomology. It is quite straight forward to see what kind of complex it comes from. It is called the *de Rham complex* and is given as

$$\cdots \longrightarrow \Omega^0(\mathcal{M}) \xrightarrow{d} \Omega^1(\mathcal{M}) \xrightarrow{d} \Omega^2(\mathcal{M}) \xrightarrow{d} \Omega^3(\mathcal{M}) \xrightarrow{d} \cdots. \qquad (4.14)$$

## 4.2. Cohomology: The idea

Since we know that the differential d satisfies $d^2 = 0$, this is indeed a complex where the cohomology is the de Rham cohomology

$$H_{\mathrm{dR}}^i(\mathcal{M}) = \frac{\ker(\mathrm{d})}{\mathrm{im}(\mathrm{d})}, \qquad (4.15)$$

where the d here is the corresponding mapping $\mathrm{d}: \Omega^i \longrightarrow \Omega^{i+1}$.

### 4.2.2 Dolbeault cohomology

The concept of de Rham cohomology is valid for all smooth manifolds. Nevertheless sometimes we do have some more structure on our manifold that allows us to derive a more refined version of de Rham cohomology. Especially when we have a complex manifold which means that we have a complex structure that tell us how complex conjugation has to be performed and ultimately every patch in our open cover can be represented locally as $\mathbb{C}$. It follows that the differential d defined in (4.2) now decomposes with respect to the complex structure of the complex manifold as

$$\mathrm{d} = \partial + \bar{\partial} := \sum_i \mathrm{d}z_i \wedge \frac{\partial}{\partial z_i} + \sum_i \mathrm{d}\bar{z}_i \wedge \frac{\partial}{\partial \bar{z}_i} \qquad (4.16)$$

Hence we can assign now two indices $(p, q)$ to differential forms where the first denotes the $\partial$-degree and the second the $\bar{\partial}$-degree.

**Definition 4.2.3** (Dolbeault complex/cohomology, Hodge diamond). Making use of this idea, we can treat both operators $\partial$ and $\bar{\partial}$ separately to define cohomology. This means we have $p$-forms that can be $\partial$-closed but not $\partial$-exact along with differential forms that may be $\bar{\partial}$-closed but not $\bar{\partial}$-exact. This splits the aforementioned de Rham cohomology into two subsets which actually have no overlap. The two *Dolbeault complexes* are then given by with

$$\begin{aligned} \cdots &\longrightarrow \Omega^{0,q}(\mathcal{M}) \xrightarrow{\partial} \Omega^{1,q}(\mathcal{M}) \xrightarrow{\partial} \Omega^{2,q}(\mathcal{M}) \xrightarrow{\partial} \Omega^{3,q}(\mathcal{M}) \xrightarrow{\partial} \cdots \\ \cdots &\longrightarrow \Omega^{p,0}(\mathcal{M}) \xrightarrow{\bar{\partial}} \Omega^{p,1}(\mathcal{M}) \xrightarrow{\bar{\partial}} \Omega^{p,2}(\mathcal{M}) \xrightarrow{\bar{\partial}} \Omega^{p,3}(\mathcal{M}) \xrightarrow{\bar{\partial}} \cdots \end{aligned} \qquad (4.17)$$

and the corresponding cohomology groups $H^{p,q}(\mathcal{M})$ are given by the cohomologies of these complexes. Comparing to the de Rham cohomology we will find that the splitting of the differential forms into complex conjugate parts also applies for the cohomology and hence

$$H_{\mathrm{dR}}^k(\mathcal{M}) = \bigoplus_{p+q=k} H^{p,q}(\mathcal{M}). \qquad (4.18)$$

Sometimes people arrange the Dolbeault cohomology groups in a table which is called the *Hodge diamond*

$$
\begin{array}{ccccccccc}
 & & & & H^{0,0}_{\mathcal{M}} & & & & = H^0_{\mathcal{M}} \\
 & & & H^{1,0}_{\mathcal{M}} & \oplus & H^{0,1}_{\mathcal{M}} & & & = H^1_{\mathcal{M}} \\
 & & H^{2,0}_{\mathcal{M}} & \oplus & H^{1,1}_{\mathcal{M}} & \oplus & H^{0,2}_{\mathcal{M}} & & = H^2_{\mathcal{M}} \\
 & & & & \vdots & & & & \vdots \\
 & H^{d,d-2}_{\mathcal{M}} & \oplus & H^{d-1,d-1}_{\mathcal{M}} & \oplus & H^{d-2,d}_{\mathcal{M}} & & & = H^{2d-2}_{\mathcal{M}} \\
 & & H^{d,d-1}_{\mathcal{M}} & \oplus & H^{d-1,d}_{\mathcal{M}} & & & & = H^{2d-1}_{\mathcal{M}} \\
 & & & & H^{d,d}_{\mathcal{M}} & & & & = H^{2d}_{\mathcal{M}}
\end{array} \quad , \quad (4.19)
$$

where we used the notation

$$H^{p,q}_{\mathcal{M}} := H^{p,q}(\mathcal{M}) \quad \text{and} \quad H^k_{\mathcal{M}} := H^k_{\text{dR}}(\mathcal{M}). \tag{4.20}$$

## 4.3 Sheaf cohomology

After we have seen the quite instructive examples of de Rham and Dolbeault cohomology we want to come now to the very generic concept of sheaf cohomology and especially the cohomology of line bundles.

### 4.3.1 Čech cohomology

The most generic way to perform cohomology computations is that of Čech cohomology. It is formulated without any fancy objects such as differential forms etc. and is only making use of the open cover

$$\{U_i | i = 1, ..., n\} = \mathcal{U} \tag{4.21}$$

of the topological space $\mathbb{P}_\Sigma$ and a sheaf $\mathcal{L}$ on it. Let $\Gamma(U; \mathcal{L})$ be the set of sections in $\mathcal{L}|_U$ where $U \subset \mathcal{U}$.

**Definition 4.3.1** (Čech cochain, compelx and cohomology). A sheaf is completely determined by a set of sections, called *global sections*, and by the set of possible sections that can be defined on subsheaves of the original sheaf, we may call them *local sections* in the remainder. The cohomology of a sheaf tells us then which particular sections of the sheaf can be associated to topologically non-trivial pieces of the total space. In order to see whether or not a particular

## 4.3. Sheaf cohomology

section gives rise to such a non-trivial region of the space, we have to calculate the so-called *Čhech cochain complex* to each potential section of the sheaf. The cochains contain the information about all the possible ways to define that particular global or local section of the sheaf. Furthermore, each cochain comes with a degree which gives information about the open sets on which the section in question is defined on. Doing this for all sections, the plain definition of the $p^{\text{th}}$ cochain can be stated as follows:

$$\check{C}^p(\mathcal{U};\mathcal{L}) := \bigoplus_{i_0<...<i_p} \Gamma(\mathcal{L}|_{U_{i_1...i_p}}), \qquad (4.22)$$

where $\mathcal{L}|_{U_{i_1...i_p}} := \mathcal{L}|_{U_{i_1}\cap...\cap U_{i_p}}$. This means we consider a subsheaf $\mathcal{L}|_{U_{i_1...i_p}}$ over some subset of the toric variety which is given by the intersection of $p$ different "maximal" open sets $U_{i_1} \cap ... \cap U_{i_p}$, where the term maximal indicates that these open sets are, in our cases, corresponding to the maximal cones in the fan as introduced in chapter 3. Then the $p^{\text{th}}$ cochain contains all local sections that are defined over subsets of "depth" $p$. The $(p+1)^{\text{th}}$ cochain will therefore always contain a subset of the $p^{\text{th}}$ cochain and we can hence map the $p^{\text{th}}$ to the $(p+1)^{\text{th}}$ cochain by restricting to that subset

$$\mathrm{d}^p : \check{C}^p \longrightarrow \check{C}^{p+1} \qquad (4.23)$$

which fulfills the property

$$\mathrm{d}^p \circ \mathrm{d}^{p+1} = 0. \qquad (4.24)$$

Therefore connecting all cochains this way we arrive at a complex called the *Čech cochain complex* or simply *Čech complex*

$$\check{C}^\bullet(\mathcal{U};\mathcal{L}) : 0 \longrightarrow \check{C}^0(\mathcal{U};\mathcal{L}) \xrightarrow{\mathrm{d}^0} \check{C}^1(\mathcal{U};\mathcal{L}) \xrightarrow{\mathrm{d}^1} \check{C}^2(\mathcal{U};\mathcal{L}) \xrightarrow{\mathrm{d}^2} .... \qquad (4.25)$$

Now we can see which of these open sets that the section in question is defined on, is in fact topologically non-trivial by calculating the cohomology of the Čech complex (4.25) at position $p$, as described in section 4.2, which is then called the $p^{\text{th}}$ *Čech cohomology group* and is given by

$$H^p(\mathcal{U};\mathcal{L}) := \frac{\ker(\mathrm{d}^p)}{\mathrm{im}(\mathrm{d}^{p-1})}. \qquad (4.26)$$

There is one very important theorem that we will use all the time in the following chapters. In order to state that theorem we have to introduce one specific property of the Čech cochains.

**Theorem 4.3.1** ($\check{C}^i(\cdot)$ preserves exactness). *The operation $\check{C}^i(\cdot)$ that maps a sheaf to its Čech complex has the nice property that it preserves exactness of a sequence of sheaves. This means that if*

$$0 \longrightarrow \mathcal{R} \xhookrightarrow{G} \mathcal{L} \xtwoheadrightarrow{R} \mathcal{T} \longrightarrow 0 \tag{4.27}$$

*is an exact sequence and hence $\mathrm{im}\,(G) = \ker(R)$, then their $p^{th}$ cochains also form a short exact sequence and the following diagram commutes:*

$$\begin{array}{ccccccccc}
& & \vdots & & \vdots & & \vdots & & \\
& & \downarrow & & \downarrow & & \downarrow & & \\
0 & \longrightarrow & \check{C}^p(\mathcal{U};\mathcal{R}) & \xhookrightarrow{G} & \check{C}^p(\mathcal{U};\mathcal{L}) & \xtwoheadrightarrow{R} & \check{C}^p(\mathcal{U};\mathcal{T}) & \longrightarrow & 0 \\
& & \downarrow{d^p} & & \downarrow{d^p} & & \downarrow{d^p} & & \\
0 & \longrightarrow & \check{C}^{p+1}(\mathcal{U};\mathcal{R}) & \xhookrightarrow{G} & \check{C}^{p+1}(\mathcal{U};\mathcal{L}) & \xtwoheadrightarrow{R} & \check{C}^{p+1}(\mathcal{U};\mathcal{T}) & \longrightarrow & 0 \\
& & \downarrow & & \downarrow & & \downarrow & & \\
& & \vdots & & \vdots & & \vdots & &
\end{array} \tag{4.28}$$

With this theorem in hand we can state the following:

**Theorem 4.3.2** (Long exact sequences from short ones). *As above consider an exact sequence of sheaves*

$$0 \longrightarrow \mathcal{R} \xhookrightarrow{G} \mathcal{L} \xtwoheadrightarrow{R} \mathcal{T} \longrightarrow 0. \tag{4.29}$$

*From theorem 4.3.1 we can derive a long exact sequence in cohomology of these three sheaves as*

$$\begin{array}{c}
0 \longrightarrow H^0(\mathcal{U};\mathcal{R}) \xrightarrow{G^0} H^0(\mathcal{U};\mathcal{L}) \xrightarrow{R^0} H^0(\mathcal{U};\mathcal{T}) \xrightarrow{\delta^0} \\
\longrightarrow H^1(\mathcal{U};\mathcal{R}) \xrightarrow{G^1} H^1(\mathcal{U};\mathcal{L}) \xrightarrow{R^1} H^1(\mathcal{U};\mathcal{T}) \xrightarrow{\delta^1} \\
\longrightarrow H^2(\mathcal{U};\mathcal{R}) \xrightarrow{G^2} H^2(\mathcal{U};\mathcal{L}) \xrightarrow{R^2} H^2(\mathcal{U};\mathcal{T}) \xrightarrow{\delta^2} \ldots
\end{array} \tag{4.30}$$

*Here the maps $G^p$ and $R^p$ are induced by the corresponding ones in (4.29).*

*Construction of the coboundary maps.* The only thing we have to do is to construct the maps $\delta^p$ that connect the rows in (4.30). These maps

$$\delta^p : H^p(\mathcal{U};\mathcal{T}) \longrightarrow H^{p+1}(\mathcal{U};\mathcal{R}) \tag{4.31}$$

## 4.3. Sheaf cohomology

are usually referred to as *connecting homomorphism* or *coboundary maps*. They are constructed in the following way. Take an element $[t] \in H^i(\mathcal{U}; \mathcal{T})$ whose class has a representative in $\check{C}^p(\mathcal{U}; \mathcal{T})$. Since $R$ in (4.28) is surjective, we can find an element in $s \in \check{C}^p(\mathcal{U}; \mathcal{L})$ such that $R(s) = t$. We can then map this element $s$ to an element $\mathrm{d}^p s \in \check{C}^{p+1}(\mathcal{U}; \mathcal{L})$. Due to the fact that $t$ is an element in the cohomology, it is mapped to zero by $\mathrm{d}^p$ and due to commutativity of the diagram (4.28) we have

$$R(\mathrm{d}^p s) = 0. \tag{4.32}$$

Since also the $(p+1)^{\text{th}}$ row of (4.28) is exact, $\mathrm{d}^p s$ must be an element of the image of $G$ in this sequence. Furthermore again due to exactness $G$ is injective and hence $\mathrm{d}^p s$ is uniquely associated to an element $r'$ in $\check{C}^{p+1}(\mathcal{U}; \mathcal{R})$ which also belongs to a class $[r]$ of the corresponding cohomology group. This finalizes the map $\delta^p$ and we have a long exact sequence of cohomology groups indeed. In fact to be precise we would have to show that the coboundary maps that we just defined are actually independent of the choice $s \in R^{-1}(t)$ which can be done similarly. □

### 4.3.2 The toric Čech complex

What we now want is to explicitly calculate the cohomology of a line bundle in a toric variety. To do that we want to make use of the Čech complex. Let $\mathbb{P}_\Sigma$ be a toric variety with fan $\Sigma$. We consider the open cover that is given by the maximal cones $\sigma_i \in \Sigma$:

$$\mathcal{U}^{\cdot} = \{U_{\sigma_i} | i = 1, ..., m\} . \tag{4.33}$$

We will use this open cover to calculate the Čech cohomology of a line bundle corresponding to a formal sum of the toric divisors $\mathcal{D}_i = \{x_i = 0\} \subset \mathbb{P}_\Sigma$:

$$\mathcal{D} = \sum_{i=1}^{n} a_i \mathcal{D}_i, \quad a_i \in \mathbb{Z}. \tag{4.34}$$

To determine the Čech complex for the sheaf $\mathcal{O}_{\mathbb{P}_\Sigma}(\mathcal{D})$ we have to count (local) sections living on some intersection of elements of the open cover. Since the set of sections on an open set $U$ of this open cover is equivalent to the zeroth cohomology group on that open set (they are global sections of the corresponding subsheaf)

$$\Gamma(U; \mathcal{O}_{\mathbb{P}_\Sigma}(\mathcal{D})) = H^0(U; \mathcal{O}_{\mathbb{P}_\Sigma}(\mathcal{D})), \tag{4.35}$$

we have the following identity for the $p^{\text{th}}$ Čech cochain with respect to $\mathcal{U}$:

$$\check{C}^p(\mathcal{U}; \mathcal{O}_{\mathbb{P}_\Sigma}(\mathcal{D})) = \bigoplus_{i_0 < ... < i_p} H^0(U_{\sigma_{i_0}...\sigma_{i_p}}; \mathcal{O}_{\mathbb{P}_\Sigma}(\mathcal{D})). \tag{4.36}$$

Hence we have to successively count sections for all possible intersections of elements of $\mathcal{U}$. But these contributions are just counted by holomorphic functions in the homogeneous coordinates which is roughly speaking the reason why there is actually a grading of the space of sections and hence a grading of cohomology. It is not only a grading of the cohomology but also one of the Čech cochains themselves. This grading is now precisely coming from the $M$-lattice we introduced in 3.3.1 but we have to generate a little different objects than simply the polytope we considered there. Each lattice point $m \in M$ is associated to a specific Laurent monomial which represents a section in the sheaf as we will see below. For every such Laurent monomial we can derive Čech cochains $\check{C}^p(\mathcal{U}; \mathcal{O}_{\mathbb{P}_\Sigma}(\mathcal{D}))_m$ and hence a cochain complex

$$\check{C}^\bullet_m(\mathcal{U}; \mathcal{L}) : 0 \longrightarrow \check{C}^0_m(\mathcal{U}; \mathcal{L}) \xrightarrow{\mathrm{d}^0_m} \check{C}^1_m(\mathcal{U}; \mathcal{L}) \xrightarrow{\mathrm{d}^1_m} \check{C}^2_m(\mathcal{U}; \mathcal{L}) \xrightarrow{\mathrm{d}^2_m} \dots. \tag{4.37}$$

Since the differential $\mathrm{d}^\bullet$ defined before arose from the restriction map it is clear that it respects the grading of the cochain complex and hence gives rise to a differential $\mathrm{d}^\bullet_m$ of the graded complex. Therefore one can determine the graded cohomology which can be summed up to the full cohomology of the toric variety

$$H^p(\mathbb{P}_\Sigma; \mathcal{O}(\mathcal{D})) = \bigoplus_{m \in M} H^p_m(\mathbb{P}_\Sigma; \mathcal{O}(\mathcal{D})). \tag{4.38}$$

Due to the fact that the cohomology groups are of finite dimension, it is clear that only a finite number of lattice points in $M$ can contribute. Furthermore since it would be quite a pain to calculate the Čech complex and its cohomology for each lattice point we are going to divide the lattice $M$ into certain regions, called *chambers* in which each point gives rise to the same Čech complex and cohomology. We denote these chambers by $C_\bullet$ where the $\bullet$ stands for a string of plus and minus signs, one for each coordinate. A chamber that has a "+" at the $i^{\text{th}}$ position and a "−" at the $j^{\text{th}}$ is defined as

$$C_{...+...-...} := \{m \in M | ..., \langle m, v_i \rangle \geq -a_i, ..., \langle m, v_j \rangle < -a_j, ...\}, \tag{4.39}$$

where the $a_i$ are the coefficients of the divisor given in (4.34). So each sign stands for a certain half-space that includes the hyperplane $\langle m, v_i \rangle = -a_i$ in case of a plus sign, and excludes it in case of a minus sign. The chamber is then given by

## 4.3. Sheaf cohomology

the intersection of all half-spaces that come from the signs in the subscript of $C_\bullet$. As one can see from the definition (4.39), some of the chambers that correspond to certain sign strings may be empty for a specific choice of signs. Therefore the chamber decomposition of a sheaf $\mathcal{O}_{\mathbb{P}_\Sigma}(\mathcal{D})$ certainly depends on the coefficients $a_i$ in $\mathcal{D}$.

As mentioned earlier, each lattice point $m$ corresponds to a Laurent monomial which may define certain sections in a sheaf $\mathcal{O}_{\mathbb{P}_\Sigma}(\mathcal{D})$ restricted to an intersection of open sets of $\mathcal{U}$, i.e. global or local ones. We can actually write down the corresponding Laurent monomial in terms of the homogeneous coordinates. For simplicity let us first of all consider the sheaf $\mathcal{O}_{\mathbb{P}_\Sigma}(0)$ where all hyperplanes $l_i$ in the lattice go through the origin. Here the origin itself stands for the Laurent monomial

$$\prod_{i=1}^{n} x_i^0 = 1. \qquad (4.40)$$

Now we can move away from the origin in some direction. The exponent of a specific coordinate at such a point is then simply given by the inner product of the point with the corresponding vertex vector

$$\text{Laurent monomial}(m) := \text{LM}(m) = \prod_{i=1}^{n} x_i^{\langle m, v_i \rangle}. \qquad (4.41)$$

Twisting the sheaf to $\mathcal{O}_{\mathbb{P}_\Sigma}(\mathcal{D})$ results then in shifting the hyperplanes $l_i$. Still the hyperplane $l_i$ that belongs to a coordinate $x_i$ contains all the lattice points where the corresponding Laurent monomial has $x_i$ only with exponent 0. Now moving the hyperplane away from the origin would result in increasing or decreasing the power of $x_i$ at a fixed lattice point. The distance to the hyperplane, so to say, represents the power. Hence we get the following identification

$$\text{LM}(m) := \prod_{i=1}^{n} x_i^{a_i + \langle m, v_i \rangle}. \qquad (4.42)$$

From this one can see that the sequence of signs that correspond to a chamber tells us which coordinates are elements of the enumerator, namely those that have a plus sign and which are in the denominator, namely those that have a minus sign. For instance the Laurent monomials of the chamber $C_{...+...-...}$ from above would correspond to a set of Laurent monomials of the form

$$\text{LM}(m_\xi) = \frac{T(..., x_i, ...)}{... \cdot x_j \cdot ... \cdot W(..., x_j, ...)}, \qquad m_\xi \in C_{...+...-...}. \qquad (4.43)$$

Notice that we keep the sign of the power of each homogeneous coordinate as long as we stay in a chamber since a sign flip only happens by crossing the hyperplanes $l_i$. On the other hand, whether a Laurent monomial can at all define a section on some subsheaf $\mathcal{L}|_{U_{i_1,\ldots,i_p}}$ certainly depends on the poles the Laurent monomial carries and whether these poles are compatible with the open set $U_{i_1,\ldots,i_p}$. If so, it may give rise to a section but does not have to. Since the pole structure in one and the same chamber is the same for all lattice points, it is reasonable that every point gives rise to the same Čech complex and hence to the same cohomological contribution. Dimensionally we obtain the full cohomology of the sheaf by summing over all chambers and weight each chamber contribution by the number of lattice points therein:

$$h^i(\mathbb{P}_\Sigma; \mathcal{O}(D)) = \sum_{\text{chambers } C_\xi} \#\text{points}(C_\xi) \cdot h^i_{m_\xi}(\mathbb{P}_\Sigma; \mathcal{O}(D)), \qquad (4.44)$$

where $\{C_\xi\}$ is the set of all chambers, i.e. LMs and $m_\xi$ is some lattice point in chamber $C_\xi$.

**Example 4.3.1.** Before we move on to the actual calculation of Čech cohomology, let us first consider an example. Let $\mathbb{P}_\Sigma$ be the projective space $\mathbb{P}^2$ again. The three vectors $v_i$ are given by

$$v_1 = (1, 0), \; v_2 = (0, 1), \; v_3 = (-1, -1). \qquad (4.45)$$

Let us consider the sheaf that corresponds to the divisor

$$D = 0 \cdot D_1 + 0 \cdot D_2 + 4 \cdot D_3, \qquad (4.46)$$

which means that $(a_1, a_2, a_3) = (0, 0, 4)$. Here in two dimensions, the hyperplanes are lines and the lines $l_1$ and $l_2$ run through the origin in $(0, 1)$ and $(1, 0)$ direction respectively while $l_3$ is parallel to the line through $(1, -1)$ and the origin. Now we can perform the chamber decomposition of this sheaf in seven different chambers which is shown in figure 4.2a. As one can see there is only one chamber with a finite number of lattice points. We can also write down all the Laurent monomials as

$$\text{LM}(m) := x_1^{\langle m, (1,0) \rangle} \cdot x_2^{\langle m, (0,1) \rangle} \cdot x_1^{4 + \langle m, (-1,-1) \rangle} \qquad (4.47)$$

which you can look up in figure 4.2b. So the cohomology of this divisor can at most depend on these monomials.

Now we are ready to turn to the calculation of the cohomology of a sheaf. Everything we are interested in is actually the factor $h^p_{m_\xi}$ that belongs to each

## 4.3. Sheaf cohomology

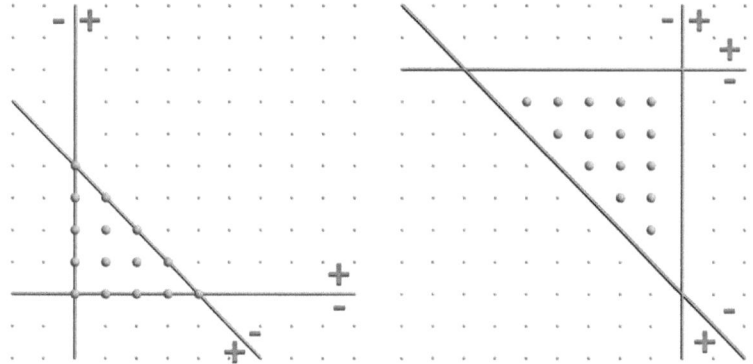

(a) Chamber decomposition of $\mathcal{O}_{\mathbb{P}^2}(4)$. The chamber $C_{+++}$ contains 15 points.

(b) Chamber decomposition of $\mathcal{O}_{\mathbb{P}^2}(-7)$. The chamber $C_{---}$ contains 15 points.

**Figure 4.2.:** The two chamber decompositions (4.39) of $\mathcal{O}_{\mathbb{P}^2}(4)$ and $\mathcal{O}_{\mathbb{P}^2}(-7)$. Every dividing line denotes power zero of the corresponding coordinate and moving towards "-" or "+" means to decrease or increase its power respectively.

chamber. As one will see, in most cases this factor is either zero or one and calculation of the full sheaf cohomology is nothing but counting Laurent monomials. But there are also subtle examples where this factor is greater than one.

**Theorem 4.3.3.** *The Laurent monomial $LM(m_\xi)$ and hence every LM of lattice points of the chamber $C_\xi$ provides a section in the sheaf $\mathcal{O}_{\mathbb{P}_\Sigma}(\mathcal{D})$ over $U_\sigma := U_{\sigma_{i_1} \ldots \sigma_{i_p}}$, if for all one-dimensional cones $\rho_i \in \sigma$ it holds that*

$$\langle m_\xi, v_i \rangle \geq -a_i, \quad \text{where } v_i \text{ is the vector that belongs to } \rho_i. \tag{4.48}$$

This is basically the condition we mentioned earlier that the degree and the pole structure of the Laurent monomial is compatible with the sheaf and the open set respectively. With this rule in hand we have everything we need to calculate the Čech cochain complex. We write down the chamber decomposition for the sheaf $\mathcal{O}_{\mathbb{P}_\Sigma}(\mathcal{D})$ in question, derive the Čech cochain complex for a representative of each chamber, derive the cohomology via the differentials and sum up all contributions in the end.

**Example 4.3.2.** Let us get back to the complex projective space $\mathbb{P}^2$ and see how we can derive the cochain complex explicitly. We already did the chamber decomposition in example 4.3.1 so let us stick with this example. Actually there is only one chamber that has a finite number of points so it is enough to derive the Čech cochain complex for this chamber because the result hast to be finite.

First of all we have the open cover

$$\mathcal{U} = \{U_1, U_2, U_3\} \tag{4.49}$$

corresponding to $\sigma_1, \sigma_2, \sigma_3$. So intersecting two an three of them leaves us with the following set of open sets:

$$\{U_1, U_2, U_3, U_{12}, U_{13}, U_{23}, U_{123}\} . \tag{4.50}$$

corresponding to the fan of $\mathbb{P}^2$. If we consider a lattice point $m$ in $C_{+++}$. This corresponds to the case where all homogeneous coordinates are in the enumerator so there is no chance that this Laurent monomial has a pole anywhere. Therefore Laurent monomials of this kind define sections on every single set in (4.50). Hence the Čech cochains have all maximal dimension and hence we can count the dimension of the cochain complex simply from the corresponding number of cones in the fan. We have three two-dimensional cones three one-dimensional cones and one zero-dimensional cone and hence obtain

$$0 \longrightarrow \check{c}_m^0 = 3 \xrightarrow{\mathrm{d}_m^0} \check{c}_m^1 = 3 \xrightarrow{\mathrm{d}_m^1} \check{c}_m^2 = 1 \longrightarrow 0, \quad \check{c}_m^i := \dim\left[\check{C}_m^i(\mathcal{U}; \mathcal{O}(4))\right] . \tag{4.51}$$

To avoid writing the same things over and over again we will in the remainder often just write down only the dimensions of the space and suppress everything else which would be here

$$0 \longrightarrow 3 \longrightarrow 3 \longrightarrow 1 \longrightarrow 0 . \tag{4.52}$$

Before we move on to calculate the cohomology we need another theorem.

**Theorem 4.3.4** (Serre duality). *Let $\mathbb{P}_\Sigma$ be a toric variety. For the cohomology of a sheaf $\mathcal{O}_{\mathbb{P}_\Sigma}(\mathcal{D})$ the following isomorphism holds:*

$$\boxed{H^p(\mathbb{P}_\Sigma; \mathcal{O}_{\mathbb{P}_\Sigma}(\mathcal{D})) \simeq H^{n-p}(\mathbb{P}_\Sigma; \mathcal{O}_{\mathbb{P}_\Sigma}(K - \mathcal{D})),} \tag{4.53}$$

*where $K = -\sum_{i=1}^n Q_i$.*

**Example 4.3.3.** Now let us do the last step and determined the cohomology for one chamber from example 4.3.1. The cochain complex was given in (4.51). In order to obtain the cohomology, we would in principle have to calculate the maps of (4.51) and see what the quotient of the kernel by the image is. Usually we are only interested in the dimension of the cohomology groups and therefore we can make use of the fact that (4.51) is a complex in order to determine this

## 4.3. Sheaf cohomology

dimension without actually writing down the maps. But here there is no unique solution if we only argue with dimensions. So here there would be no way around calculating the maps but there is a trick to avoid calculating the explicit maps in this case which is using Serre duality 4.3.4. The Serre dual to the sheaf $\mathcal{O}_{\mathbb{P}^2}(4)$ is given using the formula in 4.3.4 and reads

$$\mathcal{O}_{\mathbb{P}^2}(-1-1-1-4) = \mathcal{O}_{\mathbb{P}^2}(-7). \tag{4.54}$$

If we perform a chamber decomposition with respect to this sheaf, we we notice that there is only one chamber that is bounded containing 15 lattice points, too. This chamber is $C_{---}$ which makes it easier to calculate the cohomology. Since these Laurent monomials have all the coordinates in the denominator, we can only use it to define a section on a patch that prevents each single coordinate to vanish. Since $U_i$ is the patch where $x_i \neq 0$, the only patch that satisfies this requirement is the patch $U_{123}$. Hence the cochain complex here reads

$$0 \longrightarrow 0 \longrightarrow 0 \longrightarrow 1 \longrightarrow 0. \tag{4.55}$$

The full space $\check{C}^2(\mathcal{U}; \mathcal{O}(-7))$ is mapped to zero by $d_m^2$ and since $d_m^1$ had no image, all of it is in the cohomology. So we find

$$\dim\left[\check{C}^2(\mathcal{U}; \mathcal{O}(-7))\right] = h^2(\mathbb{P}^2; \mathcal{O}_{\mathbb{P}^2}(-7)) \stackrel{\text{Serre}}{=} h^0(\mathbb{P}^2; \mathcal{O}_{\mathbb{P}^2}(4)) \tag{4.56}$$

This is only the contribution from one Laurent monomial and in fact we have the following 15 monomials in $C_{+++}$ that all contribute in the same way:

$$\{\text{LM}(m)|m \in C_{+++}\} = \begin{aligned}&\{\ x_1^4, x_1^3 x_2^1, x_1^2 x_2^2, x_1 x_2^3, x_2^4, x_1^3 x_3, x_1^2 x_2 x_3, x_1 x_2^2 x_3, \\ & x_2^3 x_3, x_1^2 x_3^2, x_1 x_2 x_3^2, x_2^2 x_3^2, x_1 x_3^3, x_2 x_3^3, x_3^4\ \}\ . \end{aligned} \tag{4.57}$$

The final result is hence

$$\dim\left[H^\bullet(\mathbb{P}^2; \mathcal{O}(4))\right] = \left(h^0, h^1, h^2\right) = (15, 0, 0) \tag{4.58}$$

As we could see it was rather easy to calculate the cohomology of the sheaf for $\mathbb{P}^2$. Even calculating the maps would not have been such a big challenge. But as one moves to more complicated examples it happens quite soon that the dimension of the Čech cochains becomes very large and hence the mappings may become quite complicated. Here we avoided calculating the maps completely via considering the Serre dual sheaf but also this becomes impossible fairly soon and one can not only use these dimensional arguments anymore. At that point one is forced to take the maps into account which slows down the computation a lot.

That is one of the reasons why it is worth to look for a different and better algorithm to calculate cohomologies of sheaves and that is what we will present in the remainder of this chapter. The algorithm we propose will allow us to work with much more complicated spaces and still to argue dimensionally only. We achieved this by replacing the presented Čech cochain complex by a certain simplex and will show that certain combinatorial numbers we will calculate correspond to the dimension of the cohomology. This way we exclude a lot of "trivial" sections (cochains) that would usually contribute to the Čech complex, from the very beginning and we will go around the task of explicitly calculating mappings, which takes an lot more time, for a much longer time than one could in ordinary Čech cohomology. This is the reason at the end, what makes the algorithm so fast and hence powerful.

## 4.4 Sheaf-module correspondence

In order to make use of algebraic concepts, we have to reformulate the computation of sheaf cohomology on the variety $\mathbb{P}_\Sigma$ in terms of module theory of the Cox coordinate ring $S$. In fact in the last sections of this chapter we have already made use of the correspondence between sheaves and modules by making use of the explicit representatives of global and local sections. There we used them in order to get an intuition for sheaf-cohomology. In fact this correspondence can be made rigorous and we will use it extensively in the remainder.

The sheaf-module correspondence enables us to construct quasicoherent sheaves on $\mathbb{P}_\Sigma$ from any module $M$ over $S$ that is *graded* by the class group[1] $\mathrm{Cl}(\mathbb{P}_\Sigma) \cong \mathbb{Z}^{n-d}$. For the details of the construction, see e.g. §5.3 of [46]. Since we deal with line bundles, the only important observation is that the coordinate ring $S$ itself is $\mathrm{Cl}(\mathbb{P}_\Sigma)$-graded, i.e. it has a decomposition

$$S = \bigoplus_{\mathcal{D} \in \mathrm{Cl}(\mathbb{P}_\Sigma)} S_\mathcal{D}, \quad S_\mathcal{D} \cdot S_{\mathcal{D}'} \subset S_{\mathcal{D}+\mathcal{D}'} \tag{4.59}$$

and that the graded pieces $S_\mathcal{D}$ are naturally isomorphic to the sections of twisted line bundles, namely

$$S_\mathcal{D} \cong \Gamma(\mathbb{P}_\Sigma, \mathcal{O}_{\mathbb{P}_\Sigma}(\mathcal{D})). \tag{4.60}$$

This means that the *shift* $S(\mathcal{D})$ of $S$ defined by the grading $S(\mathcal{D})_{\mathcal{D}'} = S_{\mathcal{D}+\mathcal{D}'}$

---

[1] Note that we always identify Picard group and class group of $\mathbb{P}_\Sigma$, since in the smooth case all Weil divisors are already Cartier.

## 4.4. Sheaf-module correspondence

gives rise to the line bundle $\mathcal{O}_{\mathbb{P}_\Sigma}(\mathcal{D})$ and therefore sheaf cohomology should also be computable from $S(\mathcal{D})$. As it turns out, the algebraic equivalent of sheaf cohomology on a toric variety is so-called local cohomology with support on the irrelevant ideal. The reason for this lies in the fact that the map from modules to sheaves is not injective, since starting with $S$-modules $M$ that fulfill $(B_\Sigma)^l M = 0$ for $l \gg 0$ leads to trivial sheaves. Taking global sections on the sheaf side therefore in a certain way corresponds to looking at elements with support on the irrelevant ideal on the module side. Local cohomology is then defined completely analogous to sheaf cohomology as the right-derived functor of the operation of taking supports.

For the remainder we will mostly use the finer grading of the Cox coordinate ring $S$, namely the $\mathbb{Z}^n$-grading that is given by $\deg x_i = \mathbf{e}_i \in \mathbb{Z}^n$. The connection to the class group grading is given by the map

$$Q : \mathbb{Z}^n \longrightarrow \mathrm{Cl}(X) \cong \mathbb{Z}^{n-d}, \quad \mathbf{e}_i \mapsto Q \cdot \mathbf{e}_i, \qquad (4.61)$$

where the $Q$ denotes the matrix that defines the $\mathbb{C}^*$-action of the toric variety $\mathbb{P}_\Sigma$.

**Definition 4.4.1** (J-torsin submodule, local cohomology)**.** We now introduce the necessary notions and then state the relevant special case of theorem 9.5.7 in [46]. For an $S$-module $M$ and an ideal $J \subset S$ one defines the *J-torsion submodule* or submodule *supported on $J$* by

$$\Gamma_J(M) = \{y \in M \mid J^l y = 0 \text{ for some } l \in \mathbb{N}\}. \qquad (4.62)$$

The $i^{\text{th}}$ *local cohomology* module of $M$ with support on $J$ is then defined to be the module $H^i_J(M)$ obtained from any injective resolution

$$0 \to I^0 \to I^1 \to \cdots \qquad (4.63)$$

of $M$ by taking the $i^{\text{th}}$ cohomology of the subcomplex

$$0 \to \Gamma_J(I^0) \to \Gamma_J(I^1) \to \cdots. \qquad (4.64)$$

In particular, if $M$ is graded by $\mathrm{Cl}(\mathbb{P}_\Sigma)$, then also $\Gamma_J(M)$ and all $H^i_J(M)$ will inherit this grading.

**Theorem 4.4.1** (Line bundle cohomology from local cohomology)**.** *The precise* *connection between line bundle cohomology and local cohomology is then given by*[2]

$$H^i(\mathbb{P}_\Sigma, \mathcal{O}_{\mathbb{P}_\Sigma}(\mathcal{D})) \cong H^{i+1}_{B_\Sigma}(S)_\mathcal{D} \quad \text{for } i \geq 1, \mathcal{D} \in \text{Cl}(\mathbb{P}_\Sigma). \tag{4.65}$$

Furthermore, there is an exact sequence

$$0 \to H^0_{B_\Sigma}(S)_\mathcal{D} \to S_\mathcal{D} \to H^0(\mathbb{P}_\Sigma, \mathcal{O}_{\mathbb{P}_\Sigma}(\mathcal{D})) \to H^1_{B_\Sigma}(S)_\mathcal{D} \to 0, \tag{4.66}$$

which is necessary to determine the $0^{th}$ rank of sheaf cohomology. Because of (4.60) and $H^0(\mathbb{P}_\Sigma, \mathcal{O}_{\mathbb{P}_\Sigma}(\mathcal{D})) = \Gamma(\mathbb{P}_\Sigma, \mathcal{O}_{\mathbb{P}_\Sigma}(\mathcal{D}))$, the middle map is an isomorphism. Furthermore $H^0_{B_\Sigma}(S) = 0$, since $S$ has no zero divisors and so in the special case of line bundles we get

$$H^i(\mathbb{P}_\Sigma, \mathcal{O}_{\mathbb{P}_\Sigma}(\mathcal{D})) \cong H^{i+1}_{B_\Sigma}(S)_\mathcal{D} \quad \text{for } i \geq 0, \mathcal{D} \in \text{Cl}(\mathbb{P}_\Sigma). \tag{4.67}$$

The grading of the Cox ring, mentioned earlier, is also inherited by the local cohomology modules, we may write eq. (4.67) as

$$H^i(\mathbb{P}_\Sigma, \mathcal{O}_{\mathbb{P}_\Sigma}(\mathcal{D})) = \bigoplus_{\substack{\mathbf{u} \in \mathbb{Z}^n \\ Q \cdot \mathbf{u} = \mathcal{D}}} H^{i+1}_{B_\Sigma}(S)_\mathbf{u} \tag{4.68}$$

for any $\mathcal{D} \in \text{Cl}(\mathbb{P}_\Sigma)$. So the procedure would be to try and compute all $\mathbb{Z}^n$-graded pieces of local cohomology and at the end determine sheaf cohomology of $\mathcal{O}_{\mathbb{P}_\Sigma}(\mathcal{D})$ by summing up the contributions fulfilling $Q \cdot \mathbf{u} = \mathcal{D}$, which is a matrix equation over the integers solvable by standard techniques using Ehrhart polynomials. In fact, this is nothing but the Laurent monomial counting procedure of our algorithm, but we will state this later in a more precise form.

## 4.5 Commutative algebra

Before we go on and present the algorithm we have to introduce some definitions and notions from commutative algebra in order to be able to formulate and prove it. The reader who is not interested in the mathematical background and the proof of the algorithm may actually skip these sections and move on to section 4.7 right away. There are detailed instructions on how to use the algorithm explicitly, containing also a lot of examples and in fact no knowledge of section 4.5 and 4.6 is necessary to understand and use the procedure.

---

[2]The shift in the rank comes from a shift between the ordinary and the local Čech complex, see also theorem 9.5.7 in [46].

## 4.5. Commutative algebra

**Definition 4.5.1** (Simplicial complex, faces). A *simplicial complex* $\Sigma$ over a set of vertices $\{v_1, ..., v_n\}$ is defined as a collection of subsets $\Sigma = \{\sigma_1, ..., \sigma_C\}$ such that every combination of vertices $\{v_{i_1}, ..., v_{i_j}\}$ which spans a subset of an arbitrary element $\sigma_i \in \Sigma$ is itself again an element of the simplicial complex i.e.

$$\forall i \text{ s.t. } \sigma_i \in \Sigma \text{ and } \{v_{i_1}, ..., v_{i_j}\} \subset \sigma_i \Rightarrow \{v_{i_1}, ..., v_{i_j}\} \in \Sigma. \quad (4.69)$$

For simplicity we will denote a vertex set $\{v_{i_1}, ..., v_{i_j}\}$ by $\{i_1, ..., i_j\}$ and furthermore define

$$[m] := \{1, ..., m\}. \quad (4.70)$$

The elements of $\Sigma$ are called *faces* or *simplices* and their dimension is defined by the number of elements they contain decreased by one:

$$\dim(\sigma) := |\sigma| - 1 := \#\text{elements}(\sigma) - 1. \quad (4.71)$$

We also define the dimension of a simplicial complex and the dimension of the empty simplicial complex, i.e. the *void complex*, respectively as

$$\begin{aligned}\dim(\Sigma) &:= \max(\{\dim(\sigma) : \sigma \in \Sigma\}), \\ \dim(\Sigma) &:= -\infty \quad \text{for } \Sigma = \{\}.\end{aligned} \quad (4.72)$$

Notice the difference between the dimension of the empty set $\emptyset$ as an element of a simplicial complex which is

$$\dim(\emptyset) := -1 \quad \text{for } \emptyset \in \Sigma \Rightarrow \dim \Sigma = -1, \text{ for } \Sigma = \{\emptyset\} \quad (4.73)$$

and the dimension of the void complex containing no elements at all as in (4.72).

**Example 4.5.1.** An example for a simplicial complex of dimension three is shown in figure 4.3.

**Example 4.5.2.** Another example comes from the projective space $\mathbb{P}^2$ we considered in 3.2.3 and whose polytope read

$$v_1 = (-1, -1), \quad v_2 = (0, 1), \quad v_3 = (1, 0). \quad (4.74)$$

This is the vertex set for the simplicial complex and the set of subsets is given by those faces that are the bases of the cones in the fan. Here we had three kinds of

# 4. Cohomology of Line Bundles

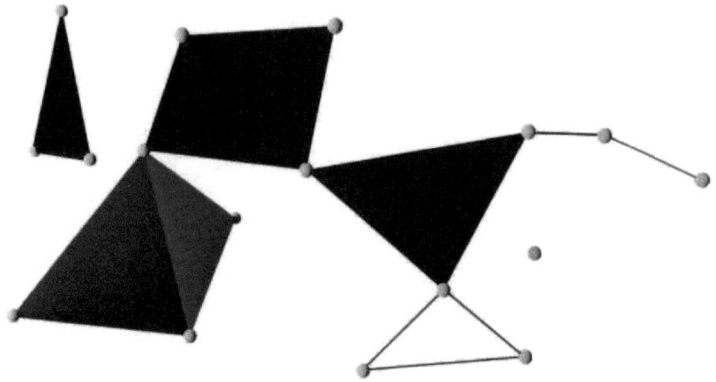

**Figure 4.3.:** The simplicial complex from example 4.5.1. It contains faces of every dimension and has maximum dimension three.

cones:

$$
\begin{array}{lll}
1 & \text{zero-dim cone } \sigma_{123} & \rightsquigarrow \quad -1\text{-dimensional face}, \\
3 & \text{one-dim cones } \sigma_{ij} & \rightsquigarrow \quad \text{zero-dimensional faces (vertices)}, \\
3 & \text{two-dim cones } \sigma_i & \rightsquigarrow \quad \text{one-dimensional faces (edges)}.
\end{array}
\quad (4.75)
$$

Since we considered the cones over faces belonging to a fan it is clear that the face that corresponds to a particular cone has one dimension less than that cone. The simplicial complex is given by

$$\Sigma = \{\emptyset, \{1\}, \{2\}, \{3\}, \{1,2\}, \{1,3\}, \{2,3\}\} . \qquad (4.76)$$

Where the $i^{\text{th}}$ vertex $i$ of the complex corresponds to the $i^{\text{th}}$ vertex $v_i$ of the polytope. Since it is clear from the definition that all subsets of a given set of simplices are included in a complex, we may suppress writing down all faces besides the maximal ones in the remainder. Here we may simply write

$$\Sigma = \{\{1,2\}, \{1,3\}, \{2,3\}\} . \qquad (4.77)$$

In fact one can derive a simplical complex as in example 4.5.2 for an arbitrary toric variety. Due to the correspondence of simplicial complexes and fans defining toric varieties, we can also give a meaning to the notions of Stanley-Reisner ideal and the irrelevant ideal for simplicial complexes from the ones we defined for toric

## 4.5. Commutative algebra

varieties. In the remainder we will continue denoting the simplicial complex with the same letter $\Sigma$ as the fan. It should be clear from the context which one we are referring to.

**Definition 4.5.2** (Stanley-Reisner and irrelevant ideal). Let $\Sigma$ be a simplicial complex over the vertex set $[n]$ and let $S = \mathbb{C}[x_1, ..., x_n]$ be the corresponding polynomial ring over $\mathbb{C}$. For some subset $\tau \in [n]$ we denote the square-free monomial in $S$ by

$$\mathbf{x}^\tau := \prod_{i \in \tau} x_i . \tag{4.78}$$

Now we define the *Stanley-Reisner ideal* of a simplicial complex $\Sigma$ to be

$$I_\Sigma := \langle \mathbf{x}^\tau | \tau \notin \Sigma \rangle . \tag{4.79}$$

Hence the generators of the Stanley-Reisner ideal are precisely made of (a minimal set of) monomials containing variables that do not span any particular simplex which agrees with definition 3.2.4.

An object, quite closely related to the Stanley-Reisner ideal just defined is the so-called *irrelevant ideal*. While we defined the SR ideal to be generated by monomial that correspond to "non-faces" of the corresponding simplex or equivalently to non-cones of the corresponding fan, the irrelevant $B_\Sigma$ ideal is defined as the ideal generated by monomials that correspond to complements of faces or cones:

$$B_\Sigma := \langle \mathbf{x}^{\overline{\sigma}} | \sigma \in \Sigma \rangle , \text{ where } \overline{\sigma} := [n] \backslash \sigma . \tag{4.80}$$

**Definition 4.5.3** (Alexander duality). The ideals that are generated by single coordinates are called *monomial prime ideals* which we define by

$$\mathfrak{m}^\tau := \langle x_i | i \in \tau \subset [n] \rangle . \tag{4.81}$$

In fact the SR ideal and the irrelevant ideal of a simplex $\Sigma$ contain the same information in the sense that there exists a duality transformation that relates both. This duality transformation is called *Alexander duality* and is defined for arbitrary monomial ideals

$$I = \langle \mathbf{x}^{\rho_1}, ..., \mathbf{x}^{\rho_m} \rangle \tag{4.82}$$

by

$$I^* := \bigcap_{j=1}^m \mathfrak{m}^{\rho_j} . \tag{4.83}$$

One can now show that

$$\boxed{I_\Sigma^* = B_\Sigma \quad \text{and} \quad B_\Sigma^* = I_\Sigma} \tag{4.84}$$

i.e. it is indeed a duality and furthermore

$$I_\Sigma := \langle \mathbf{x}^\tau | \tau \notin \Sigma \rangle = \bigcap_{\sigma \in \Sigma} \mathfrak{m}^{\overline{\sigma}} \quad \text{and} \quad B_\Sigma := \langle \mathbf{x}^{\overline{\sigma}} | \sigma \in \Sigma \rangle = \bigcap_{\tau \notin \Sigma} \mathfrak{m}^\tau . \tag{4.85}$$

Furthermore since SR and irrelevant ideal are on equal footing, there is a simplicial complex $\Sigma^*$ such that their roles interchange, i.e.

$$I_{\Sigma^*} = I_\Sigma^* = B_\sigma \quad \text{and} \quad B_{\Sigma^*} = B_\Sigma^* = I_\Sigma \tag{4.86}$$

We call $\Sigma^*$ the *Alexander dual simplicial complex* of $\Sigma$ and it is given by the complements of non-faces of the original simplex:

$$\Sigma^* = \{\overline{\tau} \mid \tau \notin \Sigma\} . \tag{4.87}$$

**Example 4.5.3** (The projective space $\mathbb{P}^2$). Let us consider a couple of examples to see how it goes. First we start with the example of $\mathbb{P}^2$ where we know that the SR ideal is given by

$$I_\Sigma = \langle x_1 x_2 x_3 \rangle . \tag{4.88}$$

Furthermore following definition (4.80), the irrelevant ideal of $\mathbb{P}^2$ is now given by:

$$\begin{aligned} B_\Sigma &= \langle \mathbf{x}^\rho \mid \overline{\rho} \in \Sigma \rangle \\ &= \langle x_1, x_2, x_3, x_1 x_2, x_1 x_3, x_2 x_3, x_1 x_2 x_3 \rangle \\ &= \langle x_1, x_2, x_3 \rangle . \end{aligned} \tag{4.89}$$

Here we saw that actually the only cones we have to consider are the maximal ones since all the non-maximal ones will give generators that are producing sub ideals and hence do not influence the ideal at all. This is always the case. Here we can already see that SR and irrelevant ideal are dual objects, but let us check explicitly that Alexander duality holds indeed which is actually trivial for the case where we have only one SR generator. From (4.85), for the SR ideal we get

$$I_\Sigma^* = \mathfrak{m}^{\{1,2,3\}} = \langle x_1, x_2, x_3 \rangle = B_\Sigma \tag{4.90}$$

## 4.5. Commutative algebra

and furthermore for the irrelevant ideal

$$B_\Sigma^* = \langle x_1 \rangle \cap \langle x_2 \rangle \cap \langle x_3 \rangle = \langle x_1 x_2 x_3 \rangle = I_\Sigma \tag{4.91}$$

which confirms Alexander duality (4.85) for this case. The Alexander dual simplex is given by the empty set $\Sigma^* = \{\emptyset\}$ which makes sense once we interpret $\langle x_1, x_2, x_3 \rangle$ as an SR ideal.

**Example 4.5.4** (The del Pezzo surface $dP_2$). In the above example we could not see the effect of performing the Alexander dual very good since the projective spaces are somewhat simple examples of toric varieties. So let us take a look at the less simple example of $dP_2$ that we have mentioned in chapter 3 already. Its fan was shown in 4.4a and its SR ideal can be read off as

$$I_\Sigma = \langle x_1 x_3, x_1 x_4, x_2 x_3, x_2 x_5, x_4 x_5 \rangle \ . \tag{4.92}$$

Our fan or simplex in this case is the collection of maximal cones:

$$\Sigma = \{\{1,2\}, \{2,4\}, \{4,3\}, \{3,5\}, \{5,1\}\} \tag{4.93}$$

The irrelevant ideal can be computed from (4.80) for $dP_2$ as:

$$B_\Sigma = \langle \mathbf{x}^{\bar\sigma} \mid \sigma \in \Sigma \rangle = \langle x_3 x_4 x_5, x_1 x_3 x_5, x_1 x_2 x_5, x_1 x_2 x_4, x_2 x_3 x_4 \rangle \ . \tag{4.94}$$

Now the Alexander dual of the SR ideal can be calculated by (4.85) to be

$$\begin{aligned} I_\Sigma^* &= \mathfrak{m}^{\{1,3\}} \cap \mathfrak{m}^{\{1,4\}} \cap \mathfrak{m}^{\{2,3\}} \cap \mathfrak{m}^{\{2,5\}} \cap, \mathfrak{m}^{\{4,5\}} \\ &= \langle x_1, x_3 \rangle \cap \langle x_1, x_4 \rangle \cap \langle x_2, x_3 \rangle \cap \langle x_2, x_5 \rangle \cap \langle x_4, x_5 \rangle \\ &= \langle x_1 x_2 x_4, x_1 x_2 x_5, x_1 x_3 x_5, x_1 x_3 x_2 x_4, x_3 x_4 x_2, x_3 x_4 x_5 \rangle \\ &= \langle x_1 x_2 x_4, x_1 x_2 x_5, x_1 x_3 x_5, x_3 x_4 x_2, x_3 x_4 x \rangle = B_\Sigma \end{aligned} \tag{4.95}$$

and the Alexander dual of the irrelevant ideal, also from (4.85), as

$$\begin{aligned} B_\Sigma^* &= \mathfrak{m}^{\{3,4,5\}} \cap \mathfrak{m}^{\{1,3,5\}} \cap \mathfrak{m}^{\{1,2,5\}} \cap \mathfrak{m}^{\{1,2,4\}} \cap, \mathfrak{m}^{\{2,3,4\}} \\ &= \langle x_3, x_4, x_5 \rangle \cap \langle x_1, x_3, x_5 \rangle \cap \langle x_1, x_2, x_5 \rangle \cap \langle x_1, x_2, x_4 \rangle \cap \langle x_2, x_3, x_4 \rangle \\ &= \langle x_3 x_1, x_3 x_2, x_4 x_1, x_4 x_5, x_5 x_2 \rangle = I_\Sigma \ . \end{aligned} \tag{4.96}$$

Hence Alexander duality also holds in this case.

**Definition 4.5.4** (Reduced (co)chain complex and (co)homology). Let $\Sigma$ be a simplicial complex on $[n]$. For each $i \geq -1$ denote by $F_i$ the set of $i$-dimensional faces as defined in 4.5.1, i.e. subsets $\sigma \subseteq [n]$ of cardinality $i+1$, and let $\mathbb{C}^{F_i}$ be the complex vector space whose basis elements $e_\sigma$ correspond to all $\sigma \in F_i$, i.e. $\mathbb{C}^{F_i} = 0$ for $i < -1$ or $i \geq n$. The *reduced chain complex* of $\Sigma$ is the complex

$$\widetilde{C}_\bullet(\Sigma): \quad 0 \longleftarrow \mathbb{C}^{F_{-1}} \xleftarrow{\partial_0} \mathbb{C}^{F_0} \xleftarrow{\partial_1} \cdots \xleftarrow{\partial_{n-1}} \mathbb{C}^{F_{n-1}} \longleftarrow 0. \quad (4.97)$$

Here the *boundary maps* $\partial_i$ are defined by setting $\mathrm{sign}(j,\sigma) = (-1)^{r-1}$ when $j$ is the $r^{\mathrm{th}}$ element of $\sigma \subseteq [n]$ written in increasing order, and

$$\partial_i(e_\sigma) = \sum_{j \in \sigma} \mathrm{sign}(j,\sigma) e_{\sigma \setminus \{j\}}. \quad (4.98)$$

One has $\partial_i \circ \partial_{i+1} = 0$ and therefore defines the $i^{\mathrm{th}}$ *reduced homology* of $\Sigma$ (with coefficients in $\mathbb{C}$) as

$$\widetilde{H}_i(\Sigma) = \frac{\ker(\partial_i)}{\mathrm{im}(\partial_{i+1})}. \quad (4.99)$$

Taking the vector space dual of the chain complex (and the transpose of all maps) one gets the *reduced cochain complex* of $\Sigma$ as $\widetilde{C}^\bullet(\Sigma) = (\widetilde{C}_\bullet(\Sigma))^*$ with *coboundary maps* $\partial^i = \partial_i^*$:

$$\widetilde{C}^\bullet(\Sigma): \quad 0 \longrightarrow \mathbb{C}^{F_{-1}*} \xrightarrow{\partial^0} \mathbb{C}^{F_0*} \xrightarrow{\partial^1} \cdots \xrightarrow{\partial^{n-1}} \mathbb{C}^{F_{n-1}*} \longrightarrow 0. \quad (4.100)$$

One similarly defines the $i^{\mathrm{th}}$ *reduced cohomology* of $\Sigma$ as

$$\widetilde{H}^i(\Sigma) = \frac{\ker(\partial^{i+1})}{\mathrm{im}(\partial^i)}. \quad (4.101)$$

Since we have coefficients in $\mathbb{C}$, there is a canonical isomorphism

$$\widetilde{H}^i(\Sigma) \cong \widetilde{H}_i(\Sigma)^* \quad (4.102)$$

and thus an equality of their dimensions

$$\dim_\mathbb{C} \widetilde{H}^i(\Sigma) = \dim_\mathbb{C} \widetilde{H}_i(\Sigma). \quad (4.103)$$

**Example 4.5.5** (Reduced (co)chain complex of $dP_2$). Let us consider the example of the del Pezzo surface $dP_2$. It is given by a blowup of $\mathbb{P}^2$ in two points. Its vertices are given as the following vectors

$$v_1 = (1,0), \ v_2 = (0,1), \ v_3 = (-1,-1), \ v_4 = (-1,0), \ v_5 = (0,-1) \quad (4.104)$$

and are shown in figure 4.4b. We want to see how the reduced (co)chain complex

## 4.5. Commutative algebra

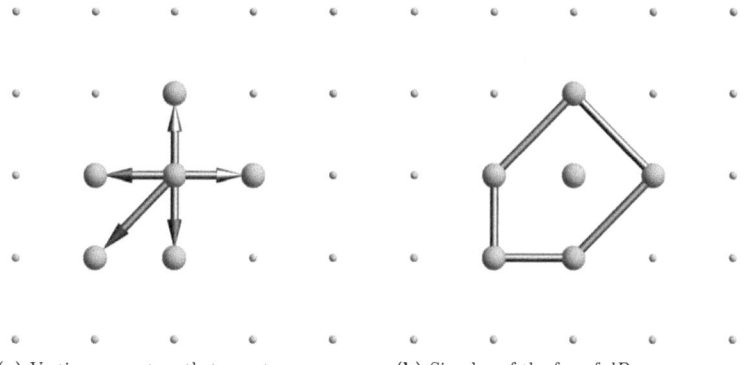

(a) Vertices as vectors that sum to zero  (b) Simplex of the fan of $dP_2$

**Figure 4.4.:** The toric diagram of $dP_2$; the vertex in the center of the image marks the origin $(0,0)$. From 4.4a we can already read off the cones in the maximal triangulation of $dP_2$

works out for the simplicial complex that corresponds to its fan. From figure 4.4 we can read off the maximal cones to be

$$\Sigma = \{\{1,2\},\{2,4\},\{4,3\},\{3,5\},\{5,1\}\}\,. \tag{4.105}$$

So we can see from 4.4b that we have five one-dimensional faces, five zero dimensional faces and as always the empty set as face of dimension minus one. Hence (4.97) becomes

$$0 \longleftarrow \mathbb{C} \xleftarrow{\partial_0} \mathbb{C}^5 \xleftarrow{\partial_1} \mathbb{C}^5 \longleftarrow 0\,. \tag{4.106}$$

The base vector corresponding to the one-dimensional face $\{1,2\} \in \Sigma$, for instance, is then mapped by the boundary map $\partial^1$ as

$$\partial_1(e_{\{1,2\}}) = -e_{\{1\}} + e_{\{2\}}\,, \tag{4.107}$$

where $\{1\}$ and $\{2\}$ are vertices of $\Sigma$. From (4.105) we can see that there is exactly one element in the kernel of $\partial^1$ which is given by

$$\ker(\partial_1) = e_{\{1,2\}} + e_{\{2,4\}} + e_{\{4,3\}} + e_{\{3,5\}} + e_{\{5,1\}}\,. \tag{4.108}$$

Since $\partial^2 = 0$ we can see that this kernel is already in the reduced homology of the complex. On the other hand, clearly the full four-dimensional image of $\partial^1$ is mapped to zero by $\partial_0$, e.g.

$$\partial_0\left(\partial_1(e_{\{1,2\}})\right) = \partial_0\left(-e_{\{1\}} + e_{\{2\}}\right) = -e_{\{\}} + e_{\{\}} = 0 \tag{4.109}$$

and there is only one linearly independent vector left that is mapped to $e_{\{\}}$ (one can choose any linear combination that is not antisymmetric in the $e_i$). So there is no more room for homology and we find

$$\widetilde{H}^1(\Sigma) = \mathbb{C} \quad \text{and} \quad \widetilde{H}^i(\Sigma) = 0, \quad \text{for} \quad i \neq 1. \tag{4.110}$$

**Definition 4.5.5** ((Minimal) free resolution). We still need some more definitions and terminology. Let $V_i, 0 \leq i \leq \ell$ be a collection of free $S$-modules, where free simply means that they are a direct sum of shifted Cox rings i.e.

$$V_i = \oplus_{q_i} S(-\mathbf{a}_{q_i}) \tag{4.111}$$

with all $\mathbf{a}_{q_i} \in \mathbb{Z}^n$ and $S(\mathbf{a})$ denoting the degree shift of $S$ by $\mathbf{a}$. A sequence

$$\mathcal{F}_\bullet: \quad 0 \longleftarrow V_0 \xleftarrow{\phi_1} V_1 \longleftarrow \cdots \longleftarrow V_{\ell-1} \xleftarrow{\phi_\ell} V_\ell \longleftarrow 0 \tag{4.112}$$

of free modules with maps fulfilling

$$\phi_i \circ \phi_{i+1} = 0 \tag{4.113}$$

is called a *complex* like we defined for vector spaces before. Such a complex is called $\mathbb{Z}^n$-*graded* if each homomorphism is of degree $\mathbf{0}$, i.e. for a homomorphism $\phi_i$ all elements $r_i \in V_i$ one has $\deg r_i = \deg \phi_i(r_i)$ in $\mathbb{Z}^n$. The *length* of the complex $\mathcal{F}_\bullet$ equals $\ell$ if $V_0 \neq 0$ and $V_\ell \neq 0$.

A complex $\mathcal{F}_\bullet$ is called a *free resolution* of an $S$-module $M$ if $\mathcal{F}_\bullet$ is *acyclic* which simply means that it is exact everywhere except in homological degree 0 and there, the homology reproduces the $S$-module i.e.

$$M = H^0(\mathcal{F}_\bullet) = V_0/\operatorname{im}(\phi_1). \tag{4.114}$$

There is a theorem from David Hilbert, called Hilbert's syzygy theorem, which says that every $S$-module has a free resolution with length at most $n$. Since we will only look at modules $M$ of the form $I$ or $S/I$, where $I$ is a monomial ideal of $S$, e.g. the SR ideal, we always get free resolutions that are naturally $\mathbb{Z}^n$-graded.

Define a partial order on $\mathbb{Z}^n$ by letting $\mathbf{a} \prec \mathbf{b}$ if the components fulfill $a_i \leq b_i$ for all $i \in [n]$. To state when a free resolution is minimal, we need to have a look

## 4.5. Commutative algebra

at the maps between the free modules in the resolution. Since the free modules are basically given by Laurent monomials monomials, the map also have to be given by matrices of such a kind. Consider the map $\phi$ as

$$\bigoplus_q S(-\mathbf{a}_q) \xleftarrow{\phi} \bigoplus_p S(-\mathbf{a}_p), \qquad (4.115)$$

between two free $S$-modules. A *monomial matrix* consists of entries $\lambda_{qp} \in \mathbb{C}$ arranged in columns labeled by the *source degrees* $\mathbf{a}_p$ and rows labeled by the *target degrees* $\mathbf{a}_q$ and whose entry $\lambda_{qp}$ is zero unless $\mathbf{a}_p \succ \mathbf{a}_q$ in the partial order of $\mathbb{Z}^n$. The map $\phi$ will then send the basis vector $\mathbf{1}_{\mathbf{a}_p}$ of $S(-\mathbf{a}_p)$ to the element $\lambda_{qp}\mathbf{x}^{\mathbf{a}_p - \mathbf{a}_q} \cdot \mathbf{1}_{\mathbf{a}_q}$ in $S(-\mathbf{a}_q)$. The condition $\mathbf{a}_p \succ \mathbf{a}_q$ then just guarantees that $\mathbf{x}^{\mathbf{a}_p - \mathbf{a}_q} \in S$ and the image of $\phi$ makes sense.

Such a monomial matrix is called *minimal* if $\lambda_{qp} = 0$ whenever $\mathbf{a}_p = \mathbf{a}_q$. Similarly, a free resolution of some module $M$ is called a *minimal free resolution* if all the maps in the resolution can be represented by minimal monomial matrices. This means that if we compare some free resolution to a minimal free resolution we would see that the ranks of the free modules $V_i$ in the latter are all simultaneously minimized compared to the former. In particular, any free resolution of $M$ contains the minimal resolution, which is unique up to isomorphisms, as a subcomplex.

**Definition 4.5.6** ((Full) Taylor resolution). As an example of a free resolution, take a monomial ideal $I = \langle m_1, \ldots, m_t \rangle$ in $S$ and write $m_\tau = \mathrm{lcm}\{m_j \mid j \in \tau\}$ for any $\tau \subseteq [t]$. Furthermore, set $\mathbf{a}_\tau = \deg(m_\tau) \in \mathbb{Z}^n$. The *full Taylor resolution* of $I$ is based on the reduced chain complex

$$\mathcal{T}_\bullet(t) := \widetilde{\mathcal{C}}_\bullet(\Sigma_{[t]}) : 0 \longleftarrow \mathbb{C}^{F_{-1}} \xleftarrow{\partial_0} \mathbb{C}^{F_0} \xleftarrow{\partial_1} \cdots \xleftarrow{\partial_{t-1}} \mathbb{C}^{F_t} \longleftarrow 0. \qquad (4.116)$$

of the full simplex $\Sigma_{[t]}$ consisting of *all* subsets of $[t]$. In contrast to those simplexes we considered so far, here we have only one maximal face consisting of all $t$ elements in the monomial ideal.

We can now define the full Taylor resolution by replacing every vector space $\mathbb{C}^{F_j}$ in $\mathcal{T}_\bullet(t)$ by the free module of the form $\oplus_{\tau \in F_j} S(-\mathbf{a}_\tau)$ and puts the boundary maps $\partial_j$ into a sequence of monomial matrices $M(\partial_j)$ with source and target labels $\mathbf{a}_\tau$ corresponding to faces $\tau \in \Sigma_{[t]}$ and entries $\lambda_{\tau, \tau \setminus \{k\}} = \mathrm{sign}(k, \tau)$ equal to the sign factors from eq. (4.98). One arrives at an acyclic complex of the form

$$\mathcal{F}_\bullet^\mathcal{T} : \quad 0 \longleftarrow S \xleftarrow{M(\partial_0)} \bigoplus_{\tau \in F_0} S(-\mathbf{a}_\tau) \xleftarrow{M(\partial_1)} \cdots \xleftarrow{M(\partial_{t-1})} S(-\mathbf{a}_{[t]}) \longleftarrow 0, \qquad (4.117)$$

whose $0^{\text{th}}$ homology equals $S/I$, so this is a free resolution of $S/I$ of length $t$. For applications, we will use this to derive the full Taylor resolution of $S/I_\Sigma$ where $I_\Sigma$ is the SR ideal of the toric variety corresponding to $\Sigma$ and hence to get it, we only need the full power set of the simplex corresponding to the SR ideal.

*Remark.* Unfortunately, the Taylor resolution is almost never minimal. More precisely, one can show that the Taylor resolution is minimal if and only if for all faces $\sigma \in \Sigma_{[t]}$ and all elements $i \in \sigma$, the monomials $m_\sigma$ and $m_{\sigma \setminus \{i\}}$ are different.[3]

**Example 4.5.6** (Taylor resolution of $dP_2$). Coming back to our example of the del Pezzo surface, we notice that its SR ideal (4.92) contains five elements and hence describes a simplex containing all subsets of [5]. Hence the reduced chain complex of this simplex is given by

$$0 \longleftarrow \mathbb{C}^1 \longleftarrow \mathbb{C}^5 \longleftarrow \mathbb{C}^{10} \longleftarrow \mathbb{C}^{10} \longleftarrow \mathbb{C}^5 \longleftarrow \mathbb{C}^1 \longleftarrow 0 \,. \tag{4.118}$$

And reading of the degrees from table 4.1 we find the full Taylor resolution to

| vertices of the polyhedron / fan | coords | GLSM charges $Q^1$ $Q^2$ $Q^3$ | divisor class |
|---|---|---|---|
| $\nu_1 = (\phantom{-}1,\phantom{-}0\,)$ | $x_1$ | 1   0   1 | $H+Y$ |
| $\nu_2 = (\phantom{-}0,\phantom{-}1\,)$ | $x_2$ | 1   1   0 | $H+X$ |
| $\nu_3 = (-1,-1\,)$ | $x_3$ | 1   0   0 | $H$ |
| $\nu_4 = (-1,\phantom{-}0\,)$ | $x_4$ | 0   0   1 | $Y$ |
| $\nu_5 = (\phantom{-}0,-1\,)$ | $x_5$ | 0   1   0 | $X$ |

$$I_\Sigma = \langle x_1 x_3, x_1 x_4, x_2 x_3, x_2 x_5, x_4 x_5 \rangle \,.$$

**Table 4.1.:** Toric data for the del Pezzo-2 surface

be

$$0 \longleftarrow \mathcal{F}^\mathcal{T}_{-1} \longleftarrow \mathcal{F}^\mathcal{T}_0 \longleftarrow \mathcal{F}^\mathcal{T}_1 \longleftarrow \mathcal{F}^\mathcal{T}_2 \longleftarrow \mathcal{F}^\mathcal{T}_3 \longleftarrow \mathcal{F}^\mathcal{T}_4 , \longleftarrow 0 \tag{4.119}$$

---
[3] For example, the Taylor resolution of the Stanley-Reisner ring of $\mathbb{P}_\Sigma = dP_3$ is not minimal, since the subset $\{m_1, m_2, m_3\} = \{x_1 x_2, x_1 x_3, x_2 x_3\}$ is among the generators of its Stanley-Reisner ideal, cf. the examples in [66].

## 4.5. Commutative algebra

where

$$\mathcal{F}_{-1}^T = S$$

$$\mathcal{F}_0^T = S\begin{pmatrix}2\\0\\1\end{pmatrix} \oplus S\begin{pmatrix}2\\1\\0\end{pmatrix} \oplus S\begin{pmatrix}1\\0\\2\end{pmatrix} \oplus S\begin{pmatrix}1\\2\\0\end{pmatrix} \oplus S\begin{pmatrix}0\\1\\1\end{pmatrix}$$

$$\mathcal{F}_1^T = S\begin{pmatrix}3\\1\\1\end{pmatrix} \oplus S\begin{pmatrix}2\\0\\2\end{pmatrix} \oplus S\begin{pmatrix}3\\2\\1\end{pmatrix} \oplus S\begin{pmatrix}2\\1\\2\end{pmatrix} \oplus S\begin{pmatrix}3\\1\\2\end{pmatrix} \oplus$$

$$S\begin{pmatrix}2\\2\\0\end{pmatrix} \oplus S\begin{pmatrix}2\\2\\1\end{pmatrix} \oplus S\begin{pmatrix}2\\2\\2\end{pmatrix} \oplus S\begin{pmatrix}1\\1\\2\end{pmatrix} \oplus S\begin{pmatrix}1\\2\\1\end{pmatrix} \quad (4.120)$$

$$\mathcal{F}_2^T = S\begin{pmatrix}3\\1\\2\end{pmatrix} \oplus S\begin{pmatrix}3\\2\\1\end{pmatrix} \oplus S\begin{pmatrix}3\\2\\2\end{pmatrix}^{\oplus 5} \oplus S\begin{pmatrix}2\\1\\2\end{pmatrix} \oplus S\begin{pmatrix}2\\2\\1\end{pmatrix} \oplus S\begin{pmatrix}2\\2\\2\end{pmatrix}$$

$$\mathcal{F}_3^T = S\begin{pmatrix}3\\2\\2\end{pmatrix}^{\oplus 5}$$

$$\mathcal{F}_4^T = S\begin{pmatrix}3\\2\\2\end{pmatrix}$$

**Definition 4.5.7** (Betti numbers). Basically all information about the minimal free resolution of an $S$-module can be encoded in a collection of integer numbers. If we take the complex $\mathcal{F}_\bullet$ from (4.112) to be a minimal free resolution of an $S$-module $M$ and write the $V_i$ as

$$V_i = \bigoplus_{\mathbf{a} \in \mathbb{Z}^n} S(-\mathbf{a})^{\beta_{i,\mathbf{a}}}, \quad (4.121)$$

then the $\beta_{i,\mathbf{a}}(M)$ is called the $i^{\text{th}}$ *Betti number* of $M$ in degree $\mathbf{a}$.

Betti numbers can also be characterized more categorically in terms of the Tor-functor[4]. For two $S$-modules $M$ and $N$ one can describe the modules $\text{Tor}_i^S(M, N)$ by applying the functor $\_\_ \otimes_S N$ to a free resolution of $M$ and taking homology of the resulting complex. Since all relevant notions have a generalization to the $\mathbb{Z}^n$-graded setting, also the Tor-modules can be given a natural $\mathbb{Z}^n$-grading. Intuitively speaking, the Betti numbers of $M$ then describe what survives when tensoring any free resolution with $\mathbb{C}$ and taking homology:

$$\beta_{i,\mathbf{a}}(M) = \dim_\mathbb{C}\left(\text{Tor}_i^S(M, \mathbb{C})_\mathbf{a}\right). \quad (4.122)$$

Note that the tensor product over $S$ of a shifted free module $S(-\mathbf{a})$ with the

---

[4]See [100] for more details on these categorical issues.

(a) In yellow a zero dimensional face $\sigma$ of the simplex.

(b) In yellow the link $(\sigma)$ which consists of six two-dimensional faces and their subfaces.

**Figure 4.5.:** An example of the link of a zero-dimensional face in a simplex $\Sigma$ of dimension two.

ground field $\mathbb{C} \cong S/\mathfrak{m}$ is equal to a copy of the ground field in degree $\mathbf{a} \in \mathbb{Z}^n$:

$$S(-\mathbf{a}) \otimes_S \mathbb{C} \cong \mathbb{C}(-\mathbf{a}) \qquad (4.123)$$

This means that one has an easy description of the degree $\mathbf{a}$ piece of a tensored resolution $\mathcal{F}_\bullet \otimes_S \mathbb{C}$. All copies of $S(-\mathbf{a})$ for some $\mathbf{a} \in \mathbb{Z}^n$ that were present in the resolution become one-dimensional vector spaces $\mathbb{C}(-\mathbf{a})$ and since all maps between source and target degrees with $\mathbf{a}_p \neq \mathbf{a}_q$ become zero, one can restrict to degree $\mathbf{a}$ by just looking at the subcomplex of $\mathcal{F}_\bullet \otimes_S \mathbb{C}$ made up of the spaces $\mathbb{C}(-\mathbf{a})$. These considerations will play a role in the proof of our theorem later.

**Definition 4.5.8** (Link, restriction). For any $\sigma \subseteq [n]$ the *link* of $\sigma$ inside the simplicial complex $\Sigma$ is the simplicial complex given by

$$\mathrm{link}_\Sigma(\sigma) = \{\tau \in \Sigma \,|\, \tau \cup \sigma \in \Sigma \text{ and } \tau \cap \sigma = \emptyset\}. \qquad (4.124)$$

Furthermore for each $\sigma \subseteq [n]$ define the *restriction* of a simplicial complex $\Sigma$ to $\sigma$ by

$$\Sigma|_\sigma = \{\tau \in \Sigma \,|\, \tau \subseteq \sigma\}. \qquad (4.125)$$

As an illustration for the link have a look at 4.5

The Betti numbers of a monomial ideal $I_\Sigma \subseteq S$ may be calculated in different ways. One possibility is to take the (co)homology of certain simplicial subcomplexes of the associated complex $\Sigma$ (resp. the Alexander dual $\Sigma^*$). This is described by the Hochster formulæ:

**Theorem 4.5.1** (Hochster formulæ). *For $\sigma \subseteq [n]$ write $\tilde{\sigma} \in \mathbb{Z}^n$ for the (squarefree) degree with components $\tilde{\sigma}_i = 1$ if $i \in \sigma$ and $\tilde{\sigma}_i = 0$ otherwise. Since the meaning can always be inferred from the context, we subsequently omit the tilde and write $\sigma$ also for the element in $\mathbb{Z}^n$. Treating $I_\Sigma$ and $S/I_\Sigma$ as (graded) S-modules, their Betti numbers lie only in squarefree degrees $\sigma$ and can be calculated by the Hochster formula*

$$\beta_{i-1,\sigma}(I_\Sigma) = \beta_{i,\sigma}(S/I_\Sigma) = \dim_{\mathbb{C}} \tilde{H}^{|\sigma|-i-1}(\Sigma|_\sigma). \tag{4.126}$$

*There is also a description of these Betti numbers in terms of the Alexander dual complex $\Sigma^*$. Then the dual Hochster formula states that*

$$\beta_{i,\sigma}(I_\Sigma) = \beta_{i+1,\sigma}(S/I_\Sigma) = \dim_{\mathbb{C}} \tilde{H}_{i-1}(\text{link}_{\Sigma^*}(\overline{\sigma})). \tag{4.127}$$

*Because of eq. (4.103), in all of these formulæ simplicial homology may be treated for cohomology when computing Betti numbers.*

## 4.6 Cohomology of line bundles: The algorithm

Let us get one step further towards the connection to cohomology and let us state the relation between the Betti numbers we introduced in definition 4.5.7 in the last section and local cohomology from section 4.4. For details we refer to the original paper [101]. For a vector $\mathbf{a} \in \mathbb{Z}^n$ let us define the negative entries in that vector by

$$\text{neg}(\mathbf{a}) := \{i \in [n] \mid \mathbf{a}_i < 0\} \subset [n]. \tag{4.128}$$

We have motivated quite early in this chapter why Laurent monomials can be used to measure cohomology. Furthermore we found it quite intuitive that degree to which these LMs contribute only depended on the corresponding local patches on which they could be defined and hence only depended on the coordinates that were in the enumerator of the LMs. So it is in fact not so surprising to find the graded parts of local cohomology of $S$ with support on the irrelevant ideal of some toric variety to be only depending on the negative entries in their degree[5]. This means that for $\mathbf{a}, \mathbf{b} \in \mathbb{Z}^n$ with matching neg, $\text{neg}(\mathbf{a}) = \text{neg}(\mathbf{b})$ the graded parts of the local cohomology match, too:

$$H^i_{B_\Sigma}(S)_{\mathbf{a}} \cong H^i_{B_\Sigma}(S)_{\mathbf{b}}. \tag{4.129}$$

---

[5] In the fine grading we are using, every entry in the grading corresponds to a homogeneous coordinate and thinking of Laurent monomials, the negative entries just tell us which of them are inverted.

86                                                                                                    4. Cohomology of Line Bundles

Hence all relevant graded parts are given by the faces $\sigma$ of the simplex $[n]$ an hence

$$H^i_{B_\Sigma}(S) = \bigoplus_{\sigma \subset [n]} H^i_{B_\Sigma}(S)_{-\sigma} \qquad (4.130)$$

As we have seen in section 4.4 these graded pieces of local cohomology can be use to obtain sheaf cohomology, which is where we want to arrive after all. Making use of Alexander duality allows us now to obtain a relation between these graded pieces of local cohomology and the Betti numbers of $S/I_\Sigma$ which will change the task of calculating sheaf cohomology basically to the one of finding all these Betti numbers. From [101] it follows that

$$H^i_{B_\Sigma}(S)_{-\sigma} \cong \mathrm{Tor}^S_{|\sigma|-i+1}(S/I_\Sigma, \mathbb{C})_\sigma \,, \qquad (4.131)$$

and application of (4.122) then gives

$$\dim_\mathbb{C}\left(H^i_{B_\Sigma}(S)_{-\sigma}\right) = \beta_{|\sigma|-i+1,\sigma}(S/I_\Sigma)\,. \qquad (4.132)$$

Inserting this into (4.68), we finally get a closed formula for the line bundle cohomology $h^i_{\mathbb{P}_\Sigma}(\mathcal{D}) := \dim_\mathbb{C} H^i(\mathbb{P}_\Sigma, \mathcal{O}_{\mathbb{P}_\Sigma}(\mathcal{D}))$ in terms the Betti numbers as

$$h^i_{\mathbb{P}_\Sigma}(\mathcal{O}_{\mathbb{P}_\Sigma}(\mathcal{D})) = \sum_{\substack{\mathbf{u} \in \mathbb{Z} \\ Q \cdot \mathbf{u} = \mathcal{D}}} \dim_\mathbb{C} H^{i+1}_{B_\Sigma}(S)_{-\mathrm{neg}(\mathbf{u})} = \sum_{\sigma \subseteq [n]} |(\mathcal{D},\sigma)| \cdot \beta_{|\sigma|-i,\sigma}(S/I_\Sigma)\,.$$
$$(4.133)$$

Here $|(\mathcal{D},\sigma)|$ counts the number of elements in the set

$$(\mathcal{D},\sigma) = \{\mathbf{u} \in \mathbb{Z}^n \,|\, Q \cdot \mathbf{u} = \mathcal{D}, \, \mathrm{neg}(\mathbf{u}) = \sigma\}\,, \qquad (4.134)$$

where $Q$ is the matrix that defines the $C^*$-action of the toric variety $\mathbb{P}_\Sigma$. Since $\mathbf{u}$ can be interpreted as a Laurent monomial by

$$\mathrm{LM}(\mathbf{u}) = \prod_{i=1}^n x_i^{\mathbf{u}_i}\,, \qquad (4.135)$$

the set $(\mathcal{D},\sigma)$ simply corresponds to the collection of all such Laurent monomials that are of the degree $\mathcal{D}$ and hence boils down to counting lattice points in some polytope.

*Remark.* To catch up with the notation of subsection 4.3.2, what we call $\sigma$ here corresponds precisely to what we denoted by $\{C_\xi\}$ and called chamber back there. Furthermore what we denoted by $(\mathcal{D},\sigma)$ above is nothing but the number of lattice points in the corresponding chamber.

## 4.6. Cohomology of line bundles: The algorithm

Now we have almost everything collected and put at the right place to state the algorithm of calculating line bundle cohomology which we first conjectured in [66]. Due to (4.133) we already know how to obtain line bundle cohomology from certain Betti numbers and the theorem we derived gives a very efficient method to calculate these Betti numbers. When the conjecture was formulated we did not know about the right terminology and referred to the Betti numbers as "remnant" or "secondary" cohomology. Here also a vanishing of many Betti numbers is implied and therefore, we divide our theorem into two parts, where the first part refers to the possible restriction to degrees in the power set of the Stanley-Reisner ideal by a vanishing of Betti numbers while the second gives a mathematically precise way to compute the sequences that determine the Betti numbers. But before we can state it we need one more definition.

**Definition 4.6.1** (Simplicial subcomplex). Let $\mathbb{P}_\Sigma$ be a complete simplicial smooth normal toric variety and

$$I_\Sigma = \langle m_1, ..., m_t \rangle \tag{4.136}$$

be its Stanley-Reisner ideal. Let us furthermore denote

$$\mathcal{P}(I_\Sigma) := \{\mathbf{a}_\tau \mid \tau \in [t]\} \tag{4.137}$$

to be the "square-free power-set" of SR ideal generators according to the definition of $\mathbf{a}_\tau$ in 4.5.6. Also Recall from definition 4.5.6 that the full Taylor resolution $\mathcal{F}_\bullet^T$ of $S/I_\Sigma$ is based on the Taylor complex $\mathcal{T}_\bullet(t)$ which is just the reduced chain complex of the full simplex $\Delta_{[t]}$. For some $\sigma \subseteq [n]$ define the *(relative) simplicial subcomplex*

$$\Gamma^\sigma = \{\tau \in [t] \mid \mathbf{a}_\tau = \sigma\} \tag{4.138}$$

of the full simplex $\Delta_{[t]}$. One can define maps between the sets $F_j(\Gamma^\sigma)$ of $j$-dimensional faces of this subcomplex similar to (4.98) by

$$\phi_j : F_j(\Gamma^\sigma) \longrightarrow F_{j-1}(\Gamma^\sigma), \quad e_\tau \mapsto \sum_{k \in \tau} \text{sign}(k, \tau) e_{\tau \setminus \{k\}}, \tag{4.139}$$

where $e_{\tau \setminus \{k\}} = 0$ if $\tau \setminus \{k\} \notin \Gamma^\sigma$ and $\text{sign}(k, \tau) = (-1)^{s-1}$ when $k$ is the $s^{\text{th}}$ element of $\tau \subseteq [t]$ written in increasing order. Since this is just the restriction of the boundary maps in $\mathcal{T}_\bullet(t)$, it is easy to see that this yields a well-defined complex $\widetilde{\mathcal{C}}_\bullet(\Gamma^\sigma)$ with associated (reduced relative) homology $\widetilde{H}_\bullet(\Gamma^\sigma)$.

Now we are ready to state the theorem:

**Theorem 4.6.1** (Cohomology of Line Bundles). *Let $\mathbb{P}_\Sigma$ be a toric variety as* above and $I_\Sigma$ its Stanley-Reisner Ideal. Let us furthermore consider a divisor $\mathcal{D} \in Cl(\mathbb{P}_\Sigma)$ and set $h^i_{\mathbb{P}_\Sigma}(\mathcal{D}) := \dim\left(H^i(\mathbb{P}_\Sigma, \mathcal{O}_{\mathbb{P}_\Sigma}(\mathcal{D}))\right)$.

1. $\forall\, \sigma \subset [n]$ where $\sigma \notin \mathcal{P}(\mathcal{I}_\Sigma)$ the associated Betti numbers vanish, i.e.

$$\beta_{j,\sigma}(S/I_\Sigma) = 0 \quad \forall j \geq 0\,. \tag{4.140}$$

Therefore the full contribution reduces to

$$h^i_{\mathbb{P}_\Sigma}(\mathcal{D}) = \sum_{\sigma \in \mathcal{P}(I_\Sigma)} |(\mathcal{D},\sigma)|\, \beta_{|\sigma|-i,\sigma}(S/I_\Sigma)\,. \tag{4.141}$$

2. The remaining Betti numbers $\beta_{|\sigma|-i,\sigma}(S/I_\Sigma)$ can be calculated from the degree $\sigma$ part of the full Taylor resolution. In terms of the relative subcomplex introduced above, we obtain

$$\beta_{j,\sigma}(S/I_\Sigma) = \dim \widetilde{H}_{j-1}\left(\Gamma^\sigma\right)\,. \tag{4.142}$$

*Proof of the theorem:* To get the desired Betti numbers, we can tensor the Taylor resolution of $S/I_\Sigma$ with $\mathbb{C} \cong S/\mathfrak{m}$, extract the degree $\tau$ part and take homology, see (4.122). As we have described around (4.123), the tensored resolution will just be made up of vector spaces $\mathbb{C}(-\mathbf{a})$ of degree $\mathbf{a}$ at the locations of $S(-\mathbf{a})$ in the original resolution. Considering the maps of the tensored resolution, note that all entries $\lambda_{\tau,\rho}$ with $\mathbf{a}_\tau \neq \mathbf{a}_\rho$ become zero, since they correspond to multiplication by $\mathbf{x}^{\mathbf{a}_\rho - \mathbf{a}_\tau} = 0$ in $S/\mathfrak{m}$. So we can easily extract graded parts. In particular, since we started with a Taylor complex, the restriction of the tensored resolution to its degree $\tau$ part will consist of all occurrences of $\mathbb{C}(-\mathbf{a}_\tau)$ with $\mathbf{a}_\tau = \sigma$ and therefore be equivalent to the (relative) complex $\Gamma^\sigma$ with maps as in (4.139), i.e.

$$\left(\mathcal{F}^\mathcal{T}_\bullet \otimes_S \mathbb{C}\right)_\sigma \cong \widetilde{\mathcal{C}}_{\bullet-1}(\Gamma^\sigma)\,. \tag{4.143}$$

The shift by one in the homological degree on the right hand side comes from the fact that the empty set in the Taylor complex lies in homological degree $-1$ while the corresponding free module $S$ in the full Taylor resolution of (4.117) lies in homological degree 0. To get $\mathrm{Tor}^S_r(S/I_\Sigma, \mathbb{C})_\tau$, we still have to take homology, yielding

$$\mathrm{Tor}^S_r(S/I_\Sigma, \mathbb{C})_\tau \cong \widetilde{H}_{r-1}(\Gamma^\sigma)\,, \tag{4.144}$$

which finally implies eq. (4.142) by taking dimensions. If $\tau \notin \mathcal{P}(I_\Sigma)$, the complex $\Gamma^\sigma$ is void and therefore has zero homology. This implies that the respective Betti numbers vanish and the expression for $h^i_{\mathbb{P}_\Sigma}(\mathcal{D})$ then follows from eq. (4.133). □

## 4.7 Explicit computations

In this section we want to give a quick overview of the steps that has to be performed to calculate the cohomology of line bundles on a toric variety avoiding the more involved Čech complex and by making use of the restricted Taylor resolution. One can read this part without any knowledge about the last section.

### 4.7.1 Compactified instructions

The more or less lengthy derivation of the theorem 4.6.1 in last section is at the end not crucial for the actual computation and it is possible to state the instructions without understanding the full theory behind.

#### The setup

In the following we are to calculate the cohomology for the line bundle $\mathcal{O}_{\mathbb{P}_\Sigma}(\mathcal{D})$ defined over the ambient toric variety $\mathbb{P}_\Sigma$ with homogeneous coordinates $x_i$, $i = 1, ..., n$ that carry the homogeneous multi-degree $Q_i{}^\alpha$. Denote the Stanley-Reisner ideal of $\mathbb{P}_\Sigma$ by

$$I_\Sigma = \langle \mathcal{S}_1, ..., \mathcal{S}_L \rangle , \qquad (4.145)$$

where each $\mathcal{S}_i$ corresponds to one generator of the ideal and hence is given by a square free monomial of homogeneous coordinates. This is all we need to determine the desired cohomology groups

$$H^\bullet \left( \mathbb{P}_\Sigma ; \mathcal{O}_{\mathbb{P}_\Sigma}(\mathcal{D}) \right) . \qquad (4.146)$$

For the following, define for some $\tau = \{i_1, ..., i_m\} \subset \{1, ..., n\} := [n]$

$$\mathbf{x}_\tau := \{x_{i_1}, ..., x_{i_m}\} \qquad (4.147)$$

to be the set of coordinates belonging to the index set $\tau$ and

$$\mathbf{x}^\tau := x_{i_1} \cdot ... \cdot x_{i_m} \qquad (4.148)$$

the corresponding monomial, as defined earlier.

### Determine the full contribution to cohomology

Cohomology of line bundles is basically determined by holomorphic sections in sub-line bundles of that line bundle. Such a (local) section, say $s$ can be written by a Laurent monomial, i.e. a fraction of two monomials depending on the homogeneous coordinates of the ambient space:

$$s = \frac{T\left(\mathbf{x}_{[n]\setminus\tau}\right)}{W(\mathbf{x}_\tau)} . \tag{4.149}$$

The homogeneous degree of this section has to be equal to the first Chern class of the line bundle according to the definition in chapter 3, namely

$$||s|| = ||T|| - ||W|| = \mathcal{D}. \tag{4.150}$$

Then every such section gives its contribution (possibly zero) to the full line bundle cohomology which means that we have

$$h^i(\mathbb{P}_\Sigma; \mathcal{O}_{\mathbb{P}_\Sigma}(\mathcal{D})) = \sum_{||s||=\mathcal{D}} h_s^i(\mathbb{P}_\Sigma; \mathcal{O}_{\mathbb{P}_\Sigma}(\mathcal{D})) . \tag{4.151}$$

As we have already seen in 4.3.2 these numbers are not all independent and only depend on the pole structure of the LM which we called chamber there. Hence we can simplify this to

$$h^i(\mathbb{P}_\Sigma; \mathcal{O}_{\mathbb{P}_\Sigma}(\mathcal{D})) = \sum_{\mathbf{x}_\tau} \#(\text{LM}_{\mathbf{x}_\tau}^\mathcal{D}) h_{\mathbf{x}_\tau}^i(\mathbb{P}_\Sigma; \mathcal{O}_{\mathbb{P}_\Sigma}(\mathcal{D})) , \tag{4.152}$$

where $\#(\text{LM}_{\mathbf{x}_\tau}^\mathcal{D})$ is the number of LMs that have the coordinates $\mathbf{x}_\tau$ in the denominator and whose degree is equal to $\mathcal{D}$. This is precisely what we called the chamber decomposition in 4.3.2 and as we already saw there, the hard bit is to calculate the $h_{\mathbf{x}_\tau}^i$.

### Determine the cohomological part

To determine the value $h_{\mathbf{x}_\tau}^i$ for all $\tau$ we can use a complex that is not given by the Čech complex as we did in 4.3.2. It provides the combinatorial Betti numbers that are given by the dimension of the homology of that complex, called the restricted Taylor resolution. Then the Betti numbers determine $h_{\mathbf{x}_\tau}^i$ as we will state in the following. Let

$$I_\Sigma =: \langle \mathbf{x}^{\tau_1}, ..., \mathbf{x}^{\tau_L} \rangle \quad \text{and} \quad \Delta_{I_\Sigma} := \{\mathbf{x}_{\tau_1}, ..., \mathbf{x}_{\tau_L}\} \tag{4.153}$$

## 4.7. Explicit computations

which means that $\Delta_{I_\Sigma}$ simply contains the sets of coordinates that correspond to the monomials of the SR ideal. We can now take the power set of this set

$$\mathcal{P}(\Delta_{I_\Sigma}) = \bigcup_{k=0}^{L} \mathcal{P}_k(\Delta_{I_\Sigma}), \qquad (4.154)$$

where $\mathcal{P}_k$ denotes the subset of the power set that corresponds to the disjoint union oft $k$ elements:

$$\mathbf{x}_\tau \in \mathcal{P}_k(\Delta_{I_\Sigma}) \Leftrightarrow \mathbf{x}_\rho = \mathbf{x}_{\tau_{m_1}} \cup ... \cup \mathbf{x}_{\tau_{m_k}}. \qquad (4.155)$$

By disjoint we simply mean that $\mathbf{x}_\tau$ may appear more than once in $\mathcal{P}_k$ if there are multiple ways to unite $k$ elements resulting in $\mathbf{x}_\tau$ Theorem 4.6.1 states then that

$$h^i_{\mathbf{x}_\tau} \neq 0 \quad \text{only if} \quad \mathbf{x}_\tau \in \mathcal{P}(\Delta_{I_\Sigma}) \qquad (4.156)$$

and we can drop all terms in (4.152) where this is not the case which is already quite a simplification. In order to calculate the remaining ones, we build a complex of vector spaces for $\mathbf{x}_\tau$ according to

$$\mathcal{C}^\bullet : 0 \xleftarrow{d_0} \mathbb{C}^{\# \mathbf{x}_\tau \in \mathcal{P}_0(\Delta_{I_\Sigma})} \xleftarrow{d_1} \mathbb{C}^{\# \mathbf{x}_\tau \in \mathcal{P}_1(\Delta_{I_\Sigma})} \xleftarrow{d_2} ..., \qquad (4.157)$$

i.e. where the $i^{\text{th}}$ space in the complex has dimension equal to the number of possible ways to obtain $\mathbf{x}_\tau$ from a union of elements in $\Delta_\Sigma$. The Betti numbers $\beta_{j,\tau}$ can then be extracted from the homology of this complex as

$$\beta_{\tau,j} = H_j(\mathcal{T}^\bullet) = \frac{\ker(d_i)}{\operatorname{im}(d_{i+1})}. \qquad (4.158)$$

Finally the $h^i_{\mathbf{x}_\tau}$ can be related to the Betti numbers via

$$h^i_{\mathbf{x}_\tau} = \beta_{|\tau|-i,\tau}. \qquad (4.159)$$

Hence, we have collected everything that is needed to determine the full line bundle cohomology according to equation (4.152).

### 4.7.2 Examples

**Example 4.7.1** (Čech cohomology for $\mathbb{P}^2$). As the first example let us consider the complex projective space $\mathbb{P}^2$ which we examined in section 4.3. First of all let us determine all LMs that might give any contribution at all. As a reminder

| LM | : | $0 \leftarrow$ | $\mathcal{T}^0 \leftarrow$ | $\mathcal{T}^1 \leftarrow$ | $0$, |
|---|---|---|---|---|---|
| $1$ | : | $0 \leftarrow$ | $1 \leftarrow$ | $0 \leftarrow$ | $0$, |
| $\dfrac{1}{x_1 x_2 x_3}$ | : | $0 \leftarrow$ | $0 \leftarrow$ | $1 \leftarrow$ | $0$. |

**Table 4.2.:** Sequences corresponding to the two contributing LMs of $\mathbb{P}^2$

the Stanley-Reisner Ideal of $\mathbb{P}^2$ is given by just one generator, namely

$$I_\Sigma = \langle x_1 x_2 x_3 \rangle \quad \Rightarrow \quad \Delta_{I_\Sigma} = \{\mathbf{x}_{1,2,3}\} . \tag{4.160}$$

So generating the power set of these generators is rather trivial and given by

$$P(\Delta_{I_\Sigma}) = \{\mathcal{P}_0, \mathcal{P}_1\} = \{\emptyset, \mathbf{x}_{1,2,3}\} . \tag{4.161}$$

Here we can already see that there are only two local sections that might contribute to any cohomology namely according to (4.161)

$$\frac{1}{1} \quad \text{and} \quad \frac{1}{x_1 x_2 x_3} . \tag{4.162}$$

It is also clear that the corresponding complexes that will determine the cohomology contribution have to be trivial since each element in the power set (4.161) appears exactly once. This determines the Betti numbers of these LMs uniquely by the homology of the sequences shown in table 4.2. Identifying

$$\beta_{0,\emptyset} = h_\emptyset^{0-0} \quad \text{and} \quad \beta_{1,\{1,2,3\}} = h_{\mathbf{x}_{\{1,2,3\}}}^{3-1} = h_{\mathbf{x}_{\{1,2,3\}}}^2 , \tag{4.163}$$

we can summarize the contributions of general Laurent monomials to the cohomology of an arbitrary line bundle as follows:

$$\begin{array}{ll} \text{Cotribution to} & \text{Laurent monomial} \\ H^0(\mathbb{P}^2; \mathcal{O}_{\mathbb{P}^2}(k)): & T(x_1, x_2, x_3), \\ H^1(\mathbb{P}^2; \mathcal{O}_{\mathbb{P}^2}(k)): & 0, \\ H^2(\mathbb{P}^2; \mathcal{O}_{\mathbb{P}^2}(k)): & \dfrac{1}{x_1 x_2 x_3 \cdot W(x_1, x_2, x_3)}, \end{array} \tag{4.164}$$

where $T$ and $W$ are monomials of degree $k$ and $k+3$ respectively. If we want to calculate for instance $\mathcal{O}_{\mathbb{P}^2}(-4)$ we see that only the second LM $\frac{1}{x_1 x_2 x_3}$ will give a

## 4.7. Explicit computations

contribution since it is the only way to produce a negative degree. Hence we find

$$H^\bullet(\mathbb{P}^2; \mathcal{O}_{\mathbb{P}^2}(-4)) = \left(0, 0, \left\langle \frac{1}{x_1^2 x_2 x_3}, \frac{1}{x_1 x_2^2 x_3}, \frac{1}{x_1 x_2 x_3^2} \right\rangle \right) \qquad (4.165)$$
$$h^\bullet(\mathbb{P}^2; \mathcal{O}_{\mathbb{P}^2}(-4)) = (0, 0, 3) \,.$$

**Example 4.7.2** (del Pezzo-1 surface)**.** As an example not as trivial as the projective space, we have a look at a complete generic line bundle over $dP_1$ which has the toric data given in table 4.3. Forming the power set according to its

| vertices of the polyhedron / fan | coords | GLSM charges $Q^1$ | $Q^2$ | divisor class |
|---|---|---|---|---|
| $\nu_1 = (\phantom{-}1,\phantom{-}0)$ | $x_1$ | 1 | 0 | $H$ |
| $\nu_2 = (\phantom{-}0,\phantom{-}1)$ | $x_2$ | 1 | 1 | $H+X$ |
| $\nu_3 = (-1, -1)$ | $x_3$ | 1 | 0 | $H$ |
| $\nu_4 = (\phantom{-}0, -1)$ | $x_4$ | 0 | 1 | $X$ |

$$\text{intersection form:} \quad HX - X^2$$
$$I_\Sigma(dP_1) = \langle x_1 x_3, \, x_2 x_4 \rangle = \langle \mathcal{S}_1, \mathcal{S}_2 \rangle$$

**Table 4.3.:** Toric data for the del Pezzo-1 surface

Stanley-Reisner ideal,

$$P(\Delta_{I_\Sigma}) = \{\mathcal{P}_0, \mathcal{P}_1, \mathcal{P}_2\} = \{\emptyset, \, \mathbf{x}_{1,3}, \, \mathbf{x}_{2,4}, \, \mathbf{x}_{1,2,3,4}\} \,, \qquad (4.166)$$

we can see that as before, every element appears again precisely once which gives again only trivial sequences as shown in table 4.4 whose non-vanishing Betti

$$\begin{array}{rccccccccc}
\text{LM:} & 0 & \longleftarrow & \mathcal{T}^0 & \longleftarrow & \mathcal{T}^1 & \longleftarrow & \mathcal{T}^2 & \longleftarrow & 0, \\
1: & 0 & \longleftarrow & 1 & \longleftarrow & 0 & \longleftarrow & 0 & \longleftarrow & 0, \\
\dfrac{1}{x_1 x_3} : & 0 & \longleftarrow & 0 & \longleftarrow & 1 & \longleftarrow & 0 & \longleftarrow & 0, \\
\dfrac{1}{x_2 x_4} : & 0 & \longleftarrow & 0 & \longleftarrow & 1 & \longleftarrow & 0 & \longleftarrow & 0, \\
\dfrac{1}{x_1 x_2 x_3 x_4} : & 0 & \longleftarrow & 0 & \longleftarrow & 0 & \longleftarrow & 1 & \longleftarrow & 0.
\end{array}$$

**Table 4.4.:** Sequences corresponding to the contributing LMs of $dP_1$

numbers are all equal to one. For the cohomology this translates to

$$\beta_{0,\emptyset} = h^0_\emptyset, \quad \beta_{1,\{1,3\}} = h^1_{\mathbf{x}_{\{1,3\}}}, \quad \beta_{1,\{2,4\}} = h^1_{\mathbf{x}_{\{2,4\}}}, \quad \beta_{2,\{1,2,3,4\}} = h^2_{\mathbf{x}_{\{1,2,3,4\}}}$$
(4.167)

which directly yields the following possible contributions to cohomology for an arbitrary line bundle $\mathcal{O}(\mathcal{D}) = \mathcal{O}(m,n)$ is given by the divisor $\mathcal{D} = mH + nX$.

$$\begin{aligned} H^0(dP_1; \mathcal{O}(m,n)) : &\quad T(x_1, x_2, x_3, x_4), \\ H^1(dP_1; \mathcal{O}(m,n)) : &\quad \frac{T(x_2, x_4)}{x_3 x_1 \cdot W(x_1, x_3)}, \quad \frac{T(x_1, x_3)}{x_2 x_4 \cdot W(x_2, x_4)}, \\ H^2(dP_1; \mathcal{O}(m,n)) : &\quad \frac{1}{x_1 x_2 x_3 x_4 \cdot W(x_1, x_2, x_3, x_4)}. \end{aligned}$$
(4.168)

Note that the rational functions representing the top cohomology class always involve the denominator monomial $\prod x_i$ of all homogeneous coordinates $x_i \in H$, which is basically Serre duality manifest in "monomial" form, since the canonical class $K = -\sum \mathcal{D}_i$ is the negative sum of all coordinate divisor classes.

In order to get the dimensions of all cohomology groups, one only has to read off the charges of the coordinates from the table to get the degrees of the respective polynomials, equate them to the degrees $(m,n)$ of the line bundle and do the remaining combinatorics, i.e. count all possible exponents. Writing $h^i(m,n)$ for the dimension of $H^i(dP_1; \mathcal{O}(m,n))$ and $\|x_i\|$ for the multi-degree of the coordinate $x_i$ in the polynomials $T$ and $W$ these steps can be schematically described as follows:

- Contributions to $h^0(m,n)$ from all combinations of exponents with

$$\deg T(x_1, x_2, x_3, x_4) = (\|x_3\| + \|x_1\| + \|x_2\|, \|x_2\| + \|x_4\|) \stackrel{!}{=} (m,n).$$
(4.169)

This gives the two different non-zero cases

1. $0 \leq m \leq n$: $\quad h^0(m,n) = \binom{m+2}{2}$.
2. $0 \leq n \leq m$: $\quad h^0(m,n) = \binom{m+2}{2} - \frac{1}{2}(m-n)(m-n+1)$.

- The degrees of the LMs contributing to $h^1(m,n)$ can be evaluated as

$$\begin{aligned} \deg \frac{T(x_2, x_4)}{x_3 x_1 \cdot W(x_1, x_3)} &= (\|x_2\| - 2 - \|x_3\| - \|x_1\|, \|x_2\| + \|x_4\|), \\ \deg \frac{T(x_1, x_3)}{x_2 x_4 \cdot W(x_2, x_4)} &= (\|x_3\| + \|x_1\| - 1 - \|x_2\|, -2 - \|x_2\| - \|x_4\|). \end{aligned}$$
(4.170)

Equating the right hand sides to $(m,n)$, this yields the cases

## 4.7. Explicit computations

1. $n \geq 0 \wedge n \geq m+2$: $\quad h^1(m,n) = \frac{1}{2}(n-m)(n-m-1)$.
2. $n \leq -2 \wedge n \leq m-1$: $\quad h^1(m,n) = \frac{1}{2}(m-n)(m-n+1)$.

- Contributions to $h^2(m,n)$ come from a LM with degree

$$\deg \frac{1}{x_1 x_2 x_3 x_4 \cdot W(x_1, x_2, x_3, x_4)} \tag{4.171}$$
$$= (-3 - \|x_3\| - \|x_1\| - \|x_2\|, -2 - \|x_2\| - \|x_4\|).$$

Setting this equal to $(m,n)$ we get

1. $n \leq m+1 \leq 0$: $\quad h^2(m,n) = \binom{-m-1}{2}$.
2. $0 \geq n \geq m+1$: $\quad h^2(m,n) = \binom{-m-1}{2} - \frac{1}{2}(n-m)(n-m-1)$.

It is easy to check that Serre duality 4.3.4 holds. Since the canonical divisor is given by the negative sum over all toric divisors of $dP_1$

$$K = -\sum_{\rho \in \Sigma(1)} D_\rho, \tag{4.172}$$

we get $K = -3H - 2X$ from the table of $dP_1$. Serre duality can then be written as

$$\begin{aligned} H^i(dP_1; \mathcal{O}(m,n)) &\cong H^{2-i}(dP_1; \mathcal{O}(K) \otimes \mathcal{O}(m,n)^\vee) \\ &\cong H^{2-i}(dP_1; \mathcal{O}(-m-3, -n-2)). \end{aligned} \tag{4.173}$$

And indeed, the computed dimensions obviously satisfy the identity

$$h^i(m,n) = h^{2-i}(-m-3, -n-2). \tag{4.174}$$

In this case the computation could still be performed just with pencil and paper but clearly for more involved higher dimensional cases a computer code is necessary. As we have explained in 4.3 is there is a correspondence of Laurent monomials and lattice points 4.42 and counting them is really reduced to counting points inside particular polytopes.

**Example 4.7.3** (del Pezzo-3 surface). Let us take two steps further and look at $dP_3$, the del Pezzo-3 surface coming from three consecutive blowups of $\mathbb{P}^2$. This is the first example we considered, for which the simplexes that determine the cohomology contributions really matter and give rise to cohomology groups of dimension two that belong to one single LM. Its toric data are given in table 4.5. First of all, the elements of the Stanley-Reisner ideal are no longer disjoint. The powerset

$$\mathcal{P}(\Delta_{I_\Sigma}) = \{\mathcal{P}_0, \mathcal{P}_1, \mathcal{P}_2, \mathcal{P}_3, \mathcal{P}_4, \mathcal{P}_5, \mathcal{P}_6, \mathcal{P}_7, \mathcal{P}_8, \mathcal{P}_9\} \tag{4.175}$$

| vertices of the polyhedron / fan | coords | GLSM charges $Q^1$ $Q^2$ $Q^3$ $Q^4$ | divisor class |
|---|---|---|---|
| $\nu_1 = (-1, \ -1)$ | $x_1$ | 1   0   0   1 | $H+Z$ |
| $\nu_2 = (\ \ 1, \ \ \ 0)$ | $x_2$ | 1   0   1   0 | $H+Y$ |
| $\nu_3 = (\ \ 0, \ \ \ 1)$ | $x_3$ | 1   1   0   0 | $H+X$ |
| $\nu_4 = (\ \ 0, \ -1)$ | $x_4$ | 0   1   0   0 | $X$ |
| $\nu_5 = (-1, \ \ \ 0)$ | $x_5$ | 0   0   1   0 | $Y$ |
| $\nu_6 = (\ \ 1, \ \ \ 1)$ | $x_6$ | 0   0   0   1 | $Z$ |

intersection form:    $HX + HY + HZ - 2H^2 - X^2 - Y^2 - Z^2$

$I_\Sigma(dP_3) = \langle x_1 x_2,\ x_1 x_3,\ x_1 x_6,\ x_2 x_3,\ x_2 x_5,\ x_3 x_4,\ x_4 x_5,\ x_4 x_6,\ x_5 x_6 \rangle$

**Table 4.5.:** Toric data for the del Pezzo-3 surface.

$$\text{LM:} \quad 0 \longleftarrow \mathcal{T}^1 \longleftarrow \mathcal{T}^2 \longleftarrow \mathcal{T}^3 \longleftarrow 0,$$
$$\frac{1}{x_1 x_2 x_3} \quad 0 \longleftarrow 0 \longleftarrow 3 \longleftarrow 1 \longleftarrow 0.$$

**Table 4.6.:** Sequences corresponding to a particular LM of $dP_3$

and we don't want to write down all the $\mathcal{P}_i$ explicitly since they contain many elements. Let us rather consider a specific element and calculate its cohomological contribution. The element $x_{\{1,2,3\}} \in \mathcal{P}$ for instance is not a unique element in the powerset but can rather be obtained in four different ways from elements of $\Delta_{I_\Sigma}$, i.e.

$$\begin{aligned}
\mathbf{x}_{\{1,2,3\}} &= \mathbf{x}_{\{1,2\}} \cup \mathbf{x}_{\{1,3\}} \in \mathcal{P}_2\,, \\
\mathbf{x}_{\{1,2,3\}} &= \mathbf{x}_{\{1,2\}} \cup \mathbf{x}_{\{2,3\}} \in \mathcal{P}_2\,, \\
\mathbf{x}_{\{1,2,3\}} &= \mathbf{x}_{\{1,3\}} \cup \mathbf{x}_{\{2,3\}} \in \mathcal{P}_2\,, \\
\mathbf{x}_{\{1,2,3\}} &= \mathbf{x}_{\{1,2\}} \cup \mathbf{x}_{\{1,3\}} \cup \mathbf{x}_{\{2,3\}} \in \mathcal{P}_3\,.
\end{aligned} \quad (4.176)$$

This determines the induced complex which can be found in 4.6 and since it is a complex it has to have a two-dimensional homology group at position two which due to shift between the cohomology and the Betti numbers results in

$$2 = \beta_{2, \mathbf{x}_{1,2,3}} = h_{\mathbf{x}_{1,2,3}}^{|\{1,2,3\}|-2} = h_{\mathbf{x}_{1,2,3}}^1\,. \quad (4.177)$$

Hence, we will find two cochains that will be in the first cohomology group of the corresponding Čech cochain complex. So for the LM $\frac{1}{x_1 x_2 x_3}$ we get a two dimensional space which contributes for instance to the line bundle cohomology

## 4.7. Explicit computations

$\mathcal{O}_{dP_3}(-3,-1,-1,-1)$

$$\dim\left[H^1_{\mathbf{x}_{1,2,3}}(dP_3;\mathcal{O}_{dP_3}(-3,-1,-1,-1))\right]=2\,. \tag{4.178}$$

In terms of cochains this roughly means that the local section $\frac{1}{x_1x_2x_3}$ can be defined in two different non-trivial ways that both lie in the cohomology of the Čech cochain complex. In more subtle examples which means in examples with extraordinary Stanley-Reisner ideals we could also find that even cases appear where one single LM contributes a space of dimension three to the cohomology.

# Chapter 5

# Equivariant Cohomology

Toric varieties as we introduced them in chapter 3 sometimes allow for a discrete group action, e.g. a $\mathbb{Z}_n$-action, in addition to the $\mathbb{C}^*$-actions. In chapter 4 we have shown how to derive the cohomology of line bundles defined on the toric variety. Since taking a finite quotient of the toric variety is usually reflected in a discrete coordinate transformation, we want to generalize our theorem 4.6.1 to this new setting. Since we could provide for a representation of the cohomology of line bundles in terms of Laurent monomials (LMs), our attempt and our conjecture will be that one can calculate the cohomology of line bundle over the quotient space, i.e. the equivariant cohomology, by simply performing the discrete action on the corresponding LMs. We claim that only the invariant LMs under this discrete action will give rise to a contribution in cohomology of the line bundle over the quotient space. The non-invariant parts will simply be projected out. We will develop and test this conjecture using index theorems involving the holomorphic Euler character and its graded pieces, so-called *Lefschetz numbers*, where the grading is introduced by the discrete actions. Starting with discrete $\mathbb{Z}_2$-actions we will work our way up to a finite $G$-action over a toric variety.

A particularly subtle point in these considerations will be the case where the LM contributes to cohomology with a two- or three-dimensional space. One has to distinguish whether this space is itself divided into invariant and non-invariant pieces or whether it contributes consistently to one of them. As we will see, the latter one seem to be the case which leads to a conjecture on the computation of these equivariant cohomologies.

## 5.1 Physical motivation: Orientifolds in type IIA/B and heterotic orbifolds

In order to reduce $\mathcal{N} = 2$ space-time supersymmetry of a type IIA/B superstring theory on Calabi-Yau three-folds[1] down to $\mathcal{N} = 1$, one needs to consider so-called orientifolds which are $\mathbb{Z}_2$ quotients. Often, only the ingredients invariant under this symmetry survive the subsequent orientifold projection, such that the theory actually lives on the quotient space. For matter zero modes, using the usual splitting into the eigenvalues of this $\mathbb{Z}_2$-action, it is necessary to consider the invariant and anti-invariant parts of the corresponding cohomology groups.

A second important application of equivariant cohomology is found in orbifold constructions, i.e. more general $\mathbb{Z}_n$-quotients, often performed in heterotic string compactifications. Since in naive Calabi-Yau compactifications quantities like the Euler characteristic are directly tied to physical properties like e.g. the number of matter generations, orbifold constructions are often used to build spaces with suitable topological numbers. Usually one finds an abundance of "plain" spaces with huge topological invariants, whereas the phenomenologically interesting areas of the topological moduli space are sparsely populated. Orbifolds can greatly help in this aspect. For example, letting $\mathbb{Z}_5$ act freely on the quintic Calabi-Yau three-fold shows $\chi(\mathbb{P}^4[5]/\mathbb{Z}_5) = \frac{1}{5}\chi(\mathbb{P}^4[5])$.

## 5.2 Topological invariants for $\mathbb{Z}_2$ involutions

A very useful tool in complex geometry is the Riemann-Roch-Hirzebruch theorem. Given a holomorphic vector bundle $\mathcal{V}$ on some complex manifold $\mathcal{X}$ of dimension $n$, it allows to compute the Euler characteristic of this bundle via its Chern character and the Todd class of the base manifold, i.e.

$$\chi(\mathcal{X};\mathcal{V}) := \sum_{i=0}^{n}(-1)^i \dim H^i(\mathcal{X};V) \stackrel{\text{RRH}}{=} \int_{\mathcal{X}} \text{ch}(\mathcal{V})\,\text{Td}(\mathcal{X}), \qquad (5.1)$$

---

[1]In fact we have not yet introduced Calabi-Yau spaces but will do so in chapter 6, where we will show that they are easily constructed as submanifolds of toric varieties

## 5.2. Topological invariants for $\mathbb{Z}_2$ involutions

where $\mathrm{ch}(\mathcal{V})$ refers to the Chern character of $\mathcal{V}$, a polynomial expression of the Chern classes

$$\mathrm{ch}(\mathcal{V}) = \dim(\mathcal{V}) + c_1(\mathcal{V}) + \frac{c_1(\mathcal{V})^2 - c_2(\mathcal{V})}{2} \\ + \frac{c_1(\mathcal{V})^3 - 3c_1(\mathcal{V})c_2(\mathcal{V}) + 3c_3(\mathcal{V})}{6} + \ldots, \quad (5.2)$$

satisfying $\mathrm{ch}(\mathcal{V} \oplus \mathcal{W}) = \mathrm{ch}(\mathcal{V}) + \mathrm{ch}(\mathcal{W})$ as well as $\mathrm{ch}(\mathcal{V} \otimes \mathcal{W}) = \mathrm{ch}(\mathcal{V})\,\mathrm{ch}(\mathcal{W})$ and $\mathrm{Td}(\mathcal{X}) = \mathrm{Td}(T_\mathcal{X})$ is the Todd class of the base space's tangent bundle, which can for a holomorphic vector bundle also be represented by a Chern class polynomial

$$\mathrm{Td}(\mathcal{V}) = 1 + \frac{1}{2}c_1(\mathcal{V}) + \frac{1}{12}\left(c_1(\mathcal{V})^2 + c_2(\mathcal{V})\right) + \ldots . \quad (5.3)$$

Note that for line bundles the Chern character simplifies to the simple Taylor expansion

$$\mathrm{ch}(\mathcal{L}) = \mathrm{e}^{c_1(\mathcal{L})} = \sum_k \frac{c_1(\mathcal{L})^k}{k!} = 1 + c_1(\mathcal{L}) + \frac{c_1(\mathcal{L})^2}{2} + \ldots \quad (5.4)$$

that naturally truncates at the dimension of the base space, leaving only a finite number of non-zero terms in the sum.

Naturally, one would like to extend the index formula (5.1) in some way to settings subject to a symmetry action on the base space, e.g. the $\mathbb{Z}_2$ space-time involution $\Omega\sigma$ of a typical orientifold operation. The vector bundle $\mathcal{V}$ must be compatible with the $\mathbb{Z}_2$ action $\sigma$ of the orientifold involution, i.e. we require the induced mapping $\sigma^*$ to fulfill $\pi \circ \sigma^* = \sigma$ where $\pi : \mathcal{V} \longrightarrow \mathcal{X}$ is the bundle's projection mapping. Then $\sigma$ induces the splitting

$$H^i(\mathcal{X};\mathcal{V}) = H^i_+(\mathcal{X};\mathcal{V}) \oplus H^i_-(\mathcal{X};\mathcal{V}) \quad (5.5)$$

of the cohomology groups. Following a general theorem, the Euler characteristic of the orientifold's "downstairs" quotient space $\mathcal{X}/\sigma$, i.e. the invariant part of the splitting, can be expressed as

$$\chi(\mathcal{X}/\sigma; \tilde{\mathcal{V}}) = \chi_+(\mathcal{X};\mathcal{V}) = \sum_{i=0}^d (-1)^i h^i_+(\mathcal{X};\mathcal{V}) = \frac{\chi^e(\mathcal{X};\mathcal{V}) + \chi^\sigma(\mathcal{X};\mathcal{V})}{2}, \quad (5.6)$$

where $\chi^e$ and $\chi^\sigma$ are Euler characteristica associated to the two group elements of $\mathbb{Z}_2 = \{e, \sigma\}$. Here $\tilde{\mathcal{V}}$ corresponds to the bundle $\mathcal{V}$ on the quotient space $\mathcal{X}/\sigma$. Since $e$ is the unit element, $\chi^e$ actually corresponds to the ordinary Euler characteristic

$$\chi^e(\mathcal{X}; \mathcal{V}) = \chi(\mathcal{X}; \mathcal{V}) := \sum_{i=0}^{d}(-1)^i h^i(\mathcal{X}; \mathcal{V})$$
$$= \sum_{i=0}^{d}(-1)^i \Big(\dim H_+^i(\mathcal{X}; \mathcal{V}) + \dim H_-^i(\mathcal{X}; \mathcal{V})\Big), \quad (5.7)$$

and from the splitting on the right hand side of this equation one directly obtains

$$\chi^\sigma(\mathcal{X}; \mathcal{V}) = \sum_{i=0}^{d}(-1)^i \Big(\dim H_+^i(\mathcal{X}; \mathcal{V}) - \dim H_-^i(\mathcal{X}; \mathcal{V})\Big), \quad (5.8)$$

which gives us a sort of measure for the dimensional asymmetry of the splitting. This quantity is called the holomorphic Lefschetz number and is related to the fixpoint set of $\sigma$, i.e. to the so-called O-planes in an orientifold setting. The simple split of the cohomology groups also allows to provide the Euler characteristic of the anti-invariant part. From (5.7) and (5.8) it follows

$$\chi_-(\mathcal{X}; \mathcal{V}) = \sum_{i=0}^{d}(-1)^i h_-^i(\mathcal{X}; \mathcal{V}) = \frac{\chi^e(\mathcal{X}; \mathcal{V}) - \chi^\sigma(\mathcal{X}; \mathcal{V})}{2} \quad (5.9)$$

in obvious similarity to (5.6). Both $\chi_+(\mathcal{X}; \mathcal{V})$ and $\chi_-(\mathcal{X}; \mathcal{V})$ are used as highly nontrivial checks for the computations carried out in the next section.

Analogous to the Riemann-Roch-Hirzebruch theorem (5.1) the holomorphic Lefschetz theorem and the Atiyah-Bott theorem allow to compute the Lefschetz number via an index formula

$$\chi^\sigma(\mathcal{X}; \mathcal{V}) = \int_{\mathcal{X}^\sigma} \mathrm{ch}_\sigma(\mathcal{V}) \frac{\mathrm{Td}(T_{\mathcal{X}^\sigma})}{\mathrm{ch}_\sigma\big(\Lambda_{-1}(\bar{N}_{\mathcal{X}^\sigma})\big)}, \quad (5.10)$$

which, as mentioned before, only depends on the fixpoint set of the involution $\sigma$. In this expression the $\Lambda_{-1}(\bar{N}_{\mathcal{X}^\sigma})$ refers to the formal alternating sum of the exterior powers of the complex conjugate normal bundle of the orientifold involution fixpoint set $\mathcal{X}^\sigma \subset \mathcal{X}$, i.e.

$$\Lambda_{-1}(\bar{N}_{\mathcal{X}^\sigma}) := \sum_{i=0}^{d}(-1)^i \Lambda^i(\bar{N}_{\mathcal{X}^\sigma}). \quad (5.11)$$

In order to define the equivariant Chern character $\mathrm{ch}_\sigma(\mathcal{V})$, the vector bundle $\mathcal{V}$ is first decomposed into a direct sum of $\sigma_*$-eigenbundles, i.e. bundles $\mathcal{V}_k$ which are either invariant or anti-invariant under the induced $\sigma_*$-action.

One of the main simplifications for $\mathbb{Z}_2$-involutions derives from the fact that the induced action on the normal bundle is simply

$$\sigma_*(N_{\mathcal{X}^\sigma}) = -N_{\mathcal{X}^\sigma}. \tag{5.12}$$

For the vector bundle $\mathcal{V}$ one first decomposes it into a direct sum of eigenbundles $\mathcal{V} = \mathcal{V}_1 \oplus \cdots \oplus \mathcal{V}_m$ with eigenvalues $\rho_k = \pm 1$ and then defines

$$\mathrm{ch}_\sigma(\mathcal{V}) := \sum_{k=1}^{m} \rho_k\, \mathrm{ch}(\mathcal{V}_k). \tag{5.13}$$

Some further information on these definitions can be found in the appendix of [99] and references therein. It should be noted that the holomorphic Lefschetz theorem can be regarded as a special case of the Atiyah-Singer fixed point theorem and the index formula is also referred to as the Atiyah-Bott theorem, see §17 of [102].

## 5.3 An algorithm conjecture for $\mathbb{Z}_2$-equivariance

The algorithm for the computation of line bundle cohomologies on toric varieties which we introduced in chapter 4 provides actual representatives for the cohomology group generators in the form of local sections, i.e. Laurent Monomials (LMs), i.e. rational functions with a single monomial in the numerator and denominator, as long as only trivial multiplicities for the individual LMs are involved. Consider for example the complex projective space $\mathbb{P}^3$ and the "sign flip" involution

$$\sigma : (x_1, x_2, x_3, x_4) \mapsto (-x_1, x_2, x_3, x_4) \tag{5.14}$$

on the homogeneous coordinates of the base, which due to the projective equivalences is equivalent to the involution

$$\tau : (x_1, x_2, x_3, x_4) \mapsto (x_1, -x_2, -x_3, -x_4). \tag{5.15}$$

The fixpoint set of this involution therefore consist of two components: The divisor $\{x_1 = 0\} \cong \mathbb{P}^2$ and the isolated fixpoint $(1, 0, 0, 0)$. Due to the simplicity of the Stanley-Reisner ideal

$$I_\Sigma(\mathbb{P}^3) = \langle x_1 x_2 x_3 x_4 \rangle, \tag{5.16}$$

the contributing LMs for the computation of $h^*(\mathbb{P}^3; \mathcal{O}(k))$ are of a particularly simple form:

$$\text{for } k \geq 0: \quad \left\{ x_1^a x_2^b x_3^c x_4^d : a+b+c+d = k \right\},$$
$$\text{for } k \leq -4: \quad \left\{ \frac{1}{x_1^{a+1} x_2^{b+1} x_3^{c+1} x_4^{d+1}} : a+b+c+d = -k-4 \right\}. \tag{5.17}$$

In order to identify the overall sign each rationom picks up, one can simply apply the involution $\sigma$ to it. However, consider for example the bundle $\mathcal{O}(-5)$ and the corresponding sign under the involutions $\sigma$ and $\tau$:

$$\underbrace{\frac{1}{x_1^2 x_2 x_3 x_4}}_{\substack{\sigma \to + \\ \tau \to -}}, \underbrace{\frac{1}{x_1 x_2^2 x_3 x_4}}_{\substack{\sigma \to - \\ \tau \to +}}, \underbrace{\frac{1}{x_1 x_2 x_3^2 x_4}}_{\substack{\sigma \to - \\ \tau \to +}}, \underbrace{\frac{1}{x_1 x_2 x_3 x_4^2}}_{\substack{\sigma \to - \\ \tau \to +}} \quad \rightsquigarrow \quad \begin{array}{l} \sigma: \ (1_+, 3_-) \\ \tau: \ (3_+, 1_-) \end{array} \tag{5.18}$$

There is obviously a mismatch in the counting of signs between the two equivalent involutions of the base, which can be seen in almost all bundles $\mathcal{O}(k)$.

Ultimately, this is due to the naive application of the base involutions to the representatives of the bundle cohomology. In mathematical terms, one needs to uplift the $\mathbb{Z}_2$-action on the base to an $\mathbb{Z}_2$-action on the bundle $\mathcal{L} = \mathcal{O}(k)$, which is called an equivariant structure and makes the diagram

$$\begin{array}{ccc} \mathcal{L} & \xrightarrow{\phi_\sigma} & \mathcal{L} \\ \pi \downarrow & & \downarrow \pi \\ \mathbb{P}^3 & \xrightarrow{\sigma} & \mathbb{P}^3 \end{array} \tag{5.19}$$

commutative. More precisely, for a generic group $G$, each element $g \in G$ induces a mapping $g : \mathcal{X} \longrightarrow \mathcal{X}$ on the base geometry and has a corresponding uplift $\phi_g : \mathcal{L} \longrightarrow \mathcal{L}$ compatible with the bundle structure. This uplift defines an equivariant structure, if it preserves the group structure, i.e. if $\phi_g \circ \phi_h = \phi_{gh}$ such that the mapping is a group homomorphism.

The apparent inconsistency of (5.18) therefore stems from the false assumption that the equivalent involutions $\sigma$ and $\tau$ in the base geometry give rise to equivalent equivariant structures $\phi_\sigma$ and $\phi_\tau$ on the bundle $\mathcal{O}(k)$. For such a setting it is therefore important to specify the equivariant structure, i.e. the uplift of the base involution to the bundle, as well.

A second non-trivial aspect in the computation of equivariant cohomology comes from the non-trivial multiplicities appearing for some denominator monomials of our algorithm. One could question, if the invariant and anti-invariant monomial contributions with non-trivial multiplicities might nevertheless con-

## 5.3. An algorithm conjecture for $\mathbb{Z}_2$-equivariance

| vertices of the polyhedron / fan | coords | \multicolumn{6}{c|}{GLSM charges} | divisor class |
|---|---|---|---|---|---|---|---|---|
| | | $Q^m$ | $Q^n$ | $Q^p$ | $Q^q$ | $Q^r$ | $Q^s$ | |
| $v_1 = (-1, -1)$ | $x_1$ | 1 | 0 | 0 | 1 | 0 | 0 | $H$ |
| $v_2 = (1, 0)$ | $x_2$ | 1 | 0 | 1 | 0 | 1 | 0 | |
| $v_3 = (0, 1)$ | $x_3$ | 1 | 1 | 0 | 0 | 0 | 0 | |
| $v_4 = (0, -1)$ | $x_4$ | 0 | 1 | 0 | 0 | 1 | 0 | $E_a$ |
| $v_5 = (-1, 0)$ | $x_5$ | 0 | 0 | 1 | 0 | 0 | 0 | $E_b$ |
| $v_6 = (1, 1)$ | $x_6$ | 0 | 0 | 0 | 1 | 0 | 0 | $E_c$ |
| $v_7 = (-1, 1)$ | $x_7$ | 0 | 0 | 0 | 0 | 1 | 1 | $E_d$ |
| $v_8 = (1, -1)$ | $x_8$ | 0 | 0 | 0 | 0 | 0 | 1 | $E_e$ |

$$I_{\widetilde{dP_5}} = HE_a + HE_b - H^2 - E_e^2 - 2E_a^2 - 2E_b^2 + E_aE_e + E_bE_d - E_d^2 - E_c^2$$

$$\mathrm{SR}(\widetilde{dP_5}) = \langle x_1x_2,\ x_1x_3,\ x_1x_6,\ x_1x_7,\ x_1x_8,\ x_2x_3,\ x_2x_4,$$
$$x_2x_5,\ x_2x_7,\ x_3x_4,\ x_3x_5,\ x_3x_8,\ x_4x_5,\ x_4x_6,$$
$$x_4x_7,\ x_5x_6,\ x_5x_8,\ x_6x_7,\ x_6x_8,\ x_7x_8 \rangle$$

**Table 5.1.:** Toric data for the non-generic $\widetilde{dP_5}$ surface, which arises via two additional blowups from the standard $dP_3$ and differs from the standard $dP_5 = \mathbb{P}^5[2,2]$.

tribute unconventionally to the invariant and anti-invariant cohomology groups. As a highly non-trivial check for this issue, we consider the non-standard del Pezzo-5 surface, which has a toric description similar to $dP_1$, $dP_2$ and $dP_3$. The relevant toric data is summarized in table 5.1. Due to the high number of 20 Stanley-Reisner ideal generators, this example yields 200 potentially contributing monomial denominators, where 56 of these have multiplicity 2 and two have multiplicity 3. Now, consider the involution

$$\sigma : x_1 \mapsto -x_1 \tag{5.20}$$

which is equivalent to 64 different "sign flips" due to the projective equivalences. The fixpoint set in the base can be determined to be

$$\mathcal{X}^\sigma := \mathrm{FP}_\sigma(\widetilde{dP_5}) = \{x_1 = 0\} \cup \{x_6 = 0\} \cup \{x_7 = 0\} \cup \{x_8 = 0\}, \tag{5.21}$$

giving four non-intersecting $\mathbb{P}^1$s inside the $\widetilde{dP_5}$. For the equivariant structure we use the canonical uplift of (5.20). Via a proper computation of (5.10) the resulting Lefschetz number is

$$\chi^\sigma(\widetilde{dP_5};\mathcal{O}(m,\ldots,s)) = \left(\frac{1}{4}+\frac{-m+n+p}{2}\right)+(-1)^n\left(\frac{1}{4}+\frac{m-q}{2}\right)$$
$$+(-1)^{m+n+r}\left(\frac{1}{4}+\frac{n+p-r}{2}\right) \qquad (5.22)$$
$$+(-1)^{m+n+r+s}\left(\frac{1}{4}+\frac{r-s}{2}\right)$$

which allows to check whether a multiplicity-3 local section like $\frac{1}{x_1 x_6 x_7 x_8}$ or $\frac{1}{x_2 x_3 x_4 x_5}$ entirely contributes to the invariant or anti-invariant cohomology. Likewise, we checked an abundance of other examples. The empirical data therefore leads us to pose the following:

**Conjecture 5.3.1** ($\mathbb{Z}_2$-equivariant cohomology computation). Let $\mathbb{P}_\Sigma$ be a toric variety and $\mathcal{O}_{\mathbb{P}_\Sigma}(\mathcal{D})$ some line bundle over $\mathbb{P}_\Sigma$. Let furthermore $\sigma : \mathbb{P}_\Sigma \longrightarrow \mathbb{P}_\Sigma$ be a $\mathbb{Z}_2$ involution acting on the base which induces an equivariant structure on the line bundle $\mathcal{O}_{\mathbb{P}_\Sigma}(\mathcal{D})$. The cohomology of $\mathcal{O}_{\mathbb{P}_\Sigma}(\mathcal{D})$ decomposes according to theorem 4.6.1 into a direct sum of spaces that arise according to its contributing Laurent monomials:

$$\text{Contributing LMs} = \{s_1,\ldots,s_N\}$$
$$\Rightarrow H^\bullet(\mathbb{P}_\Sigma;\mathcal{O}_{\mathbb{P}_\Sigma}(\mathcal{D})) = \bigoplus_{k=1}^{N} H^\bullet_{s_k}(\mathbb{P}_\Sigma;\mathcal{O}_{\mathbb{P}_\Sigma}(\mathcal{D})). \qquad (5.23)$$

The equivariant cohomology groups are then given by the direct sum of contributing spaces that correspond to Laurent monomials that are themselves invariant under the $\mathbb{Z}_2$ involution, i.e. if we put the $s_k$ in (5.23) in order such that the first $\tilde{k}$ are invariant and the last $N-\tilde{k}$ are anti-invariant by applying the involution on the homogeneous coordinates in the LM, we find

$$\boxed{\begin{aligned} H^\bullet(\mathbb{P}_\Sigma;\mathcal{O}_{\mathbb{P}_\Sigma}(\mathcal{D})) &= H^\bullet_+(\mathbb{P}_\Sigma;\mathcal{O}_{\mathbb{P}_\Sigma}(\mathcal{D})) \oplus H^\bullet_-(\mathbb{P}_\Sigma;\mathcal{O}_{\mathbb{P}_\Sigma}(\mathcal{D})),\text{ where} \\ H^\bullet_+(\mathbb{P}_\Sigma;\mathcal{O}_{\mathbb{P}_\Sigma}(\mathcal{D})) &= \bigoplus_{k=1}^{\tilde{k}} H^\bullet_{s_k}(\mathbb{P}_\Sigma;\mathcal{O}_{\mathbb{P}_\Sigma}(\mathcal{D}))\text{ and} \\ H^\bullet_-(\mathbb{P}_\Sigma;\mathcal{O}_{\mathbb{P}_\Sigma}(\mathcal{D})) &= \bigoplus_{k=\tilde{k}}^{N} H^\bullet_{s_k}(\mathbb{P}_\Sigma;\mathcal{O}_{\mathbb{P}_\Sigma}(\mathcal{D})) \end{aligned}} \qquad (5.24)$$

The simplicity of this (conjectured) algorithm to compute $\mathbb{Z}_2$-equivariant cohomologies ultimately stems from the fact that one can basically use the same involution mapping specified for the coordinates of the base toric variety directly on the LMs that represent the cohomology group—provided the used uplift of this mapping in the form of the equivariant structure has been specified appropriately. While we were writing up our results which we published in [70], the same conjecture was also posed and developed in the appendix of [103], where the authors focused on the computation of the Lefschetz numbers—whose computation can become somewhat involved due to the equivariant Chern characters in (5.10)—and also considered the standard examples $\mathbb{P}^1$, $\mathbb{P}^2$, $dP_1$ and $dP_3$ in detail. In the context of orientifolds we refer to their nice presentation of the $\mathbb{Z}_2$-equivariant material.

## 5.4 Invariants for finite group actions

The mathematical background presented in section 5.2 can be applied to more involved finite group actions. However, some of the aspects loose their specific clarity that the special case of the two-element group $\mathbb{Z}_2$ offers. Given a finite group

$$G = \{g_1, g_2, \ldots, g_m\} \tag{5.25}$$

of $m$ elements acting holomorphically on $\mathcal{X}$, the relation (5.6) between the Euler characteristic of the orbifold space $\mathcal{X}/G$ and the sum of the different Lefschetz numbers generalizes to

$$\chi(\mathcal{X}/G; \mathcal{V}) = \frac{1}{m} \sum_{g \in G} \chi^g(\mathcal{X}; \mathcal{V}) = \sum_{i=0}^{d} (-1)^i h_{\text{inv}}^i(\mathcal{X}; \mathcal{V}). \tag{5.26}$$

The index formula for the individual Lefschetz numbers (5.10) remains unchanged, but has to be computed separately for the individual fixpoint sets of each group element. In the decomposition of the vector bundle $\mathcal{V}$ into $g_*$-eigenbundles more general eigenvalues $\rho_k \in \mathbb{C}$ can now arise. The computation of those eigenvalues rests on the group action on the conjugated normal bundle $\bar{N}_{\mathcal{X}^g}$ of each component of the fixpoint set. Due to the decomposition

$$T_{\mathcal{X}}|_{\mathcal{X}^g} = T_{\mathcal{X}^g} \oplus N_{\mathcal{X}^g} \tag{5.27}$$

of the ambient space tangent bundle, the $g$-action on $N_{\mathcal{X}^g}$ is given by a proper decomposition of the differential mapping

$$\mathrm{d}g_p : T_p \mathcal{X} \longrightarrow T_{gp} \mathcal{X} \tag{5.28}$$

over a fixpoint $p = gp \in \mathcal{X}^g$. In order to obtain the Lefschetz numbers, this then allows for the computation of the action's eigenvalues on $\bar{N}_{\mathcal{X}^g}$ and the evaluation of the integral in (5.10).

## 5.5 Some explicit examples for finite group equivariance

**Example 5.5.1** (Example: $\mathbb{P}^2/\mathbb{Z}_3$). As an example for a generalization of the conjecture 5.3 for $\mathbb{Z}_2$ involutions, we consider the line bundle0 cohomology over the orbifold space $\mathbb{P}^2/\mathbb{Z}_3$. Here the group action of $\mathbb{Z}_3 = \{e, g_1, g_2\}$ on $\mathbb{P}^2$ is defined by the generator

$$g_1 : (x_1, x_2, x_3) \mapsto (\alpha x_1, \alpha^2 x_2, x_3) \qquad \text{for } \alpha := \sqrt[3]{1} = e^{\frac{2\pi i}{3}}. \tag{5.29}$$

Due to the projective relations between the homogeneous coordinates $x_i$ the mapping is equivalent to

$$\begin{aligned} g_1' &: (x_1, x_2, x_3) \mapsto (x_1, \alpha x_2, \alpha^2 x_3) \text{ and} \\ g_1'' &: (x_1, x_2, x_3) \mapsto (\alpha^2 x_1, x_2, \alpha x_3), \end{aligned} \tag{5.30}$$

i.e. $g_1 \sim g_1' \sim g_1''$ describe the same involution on the base space. Considering the Stanley-Reisner ideal $I_\Sigma(\mathbb{P}^2) = \langle x_1 x_2 x_3 \rangle$ this action therefore has three fixpoints

$$P_1 = (0, 0, 1), \qquad P_1' = (1, 0, 0), \qquad P_1'' = (0, 1, 0) \tag{5.31}$$

in $\mathbb{P}^2$. The second group element's involution is given by the square

$$\begin{aligned} g_2 := g_1^2 &: (x_1, x_2, x_3) \mapsto (\alpha^2 x_1, \alpha x_2, x_3) \\ g_2' &: (x_1, x_2, x_3) \mapsto (x_1, \alpha^2 x_2, \alpha x_3) \\ g_2'' &: (x_1, x_2, x_3) \mapsto (\alpha x_1, x_2, \alpha^2 x_3) \end{aligned} \tag{5.32}$$

leading to the same three fixpoints $P_2 = P_1$, $P_2' = P_1'$ and $P_2'' = P_1''$. Since for both non-trivial group elements the fixpoint sets consist of three components of maximal codimension which means the they the fixpoints are isolated, this example is particularly simple.

In order to determine the (conjugated) normal bundle's eigenspace decomposition under the induced $\mathbb{Z}_3$-action, we utilize that for fixpoints the general split (5.27) leads to the direct identification $(N_{\mathcal{X}^g})_p \cong T_p \mathcal{X}$, i.e. it suffices to compute the eigenvalues of the differentials (5.28) at the fixpoints. For the first fixpoint

## 5.5. Some explicit examples for finite group equivariance    109

$P_1 \in U_3 = \{x_3 \neq 0\} \subset \mathbb{P}^2$ we use the local chart given by

$$\phi_3 : U_3 \xrightarrow{\cong} \mathbb{C}^2$$
$$(x_1, x_2, x_3) \mapsto \left(\frac{x_1}{x_3}, \frac{x_2}{x_3}\right). \tag{5.33}$$

The involution mapping $g_1$ within this chart then takes the form

$$f_1^3 := \phi_3 \circ g_1 \circ \phi_3^{-1} : \mathbb{C}^2 \longrightarrow \mathbb{C}^2$$
$$(x, y) \mapsto (\alpha x, \alpha^2 y), \tag{5.34}$$

and the differential mapping at $\phi_3(P_1) = (0,0) \in \mathbb{C}^2$ is then easily computed to

$$d(f_1^3)_{P_1} = \begin{pmatrix} \frac{\partial f_{1,x}^3}{\partial x} & \frac{\partial f_{1,x}^3}{\partial y} \\ \frac{\partial f_{1,y}^3}{\partial x} & \frac{\partial f_{1,y}^3}{\partial y} \end{pmatrix}_{P_1} = \begin{pmatrix} \alpha & 0 \\ 0 & \alpha^2 \end{pmatrix} \tag{5.35}$$

Via $\det\left(d(f_1^3)_{P_1} - \lambda \mathbb{1}\right) = (\alpha - \lambda)(\alpha^2 - \lambda) = 0$ this leads to the eigenvalues $\lambda_1 = \alpha$ and $\lambda_2 = \alpha^2$, such that the action on the $2d_\mathbb{C}$ conjugated normal bundle induces the split into $\mathbb{Z}_3$-irreducible representations

$$\bar{N}_{P_1} \cong \bar{N}_{P_1}^{\bar{\alpha}} \oplus \bar{N}_{P_1}^{\bar{\alpha}^2} \cong \bar{N}_{P_1}^{\alpha} \oplus \bar{N}_{P_1}^{\alpha^2} \tag{5.36}$$

on the fixpoint $P_1$. The analogous computation yields the same result for all three fixpoints of both $g$ and $g^2$. Since $\dim \mathcal{X}^g = \dim \mathcal{X}^{g^2} = 0$ the expansion of the equivariant Chern character reduces to

$$\mathrm{ch}_g(\Lambda_{-1}\bar{N}_{P_1}) = \mathrm{ch}_g(\mathcal{O} - \bar{N}_{P_1} + \Lambda^2 \bar{N}_{P_1})$$
$$= \dim \mathcal{O} - (\alpha \dim \bar{N}_{P_1}^\alpha + \alpha^2 \dim \bar{N}_{P_1}^{\alpha^2}) + \alpha \cdot \alpha^2 \dim \Lambda^2 \bar{N}_{P_1} \tag{5.37}$$
$$= 1 - (\alpha + \alpha^2) + \alpha^3 = 1 - (-1) + 1 = 3.$$

Using $\mathrm{Td}(\mathcal{X}^g) = 1$ and $\mathrm{ch}_g(L) = \varrho_g(L; P) \in \mathbb{C}^*$ for each fixpoint component, it follows

$$\chi^g(\mathbb{P}^2; L) = \left[\int_{P_1} + \int_{P_1'} + \int_{P_1''}\right] \mathrm{ch}_g(L) \frac{\mathrm{Td}(\mathcal{X}^g)}{\mathrm{ch}_g(\Lambda_{-1}\bar{N}_{\mathcal{X}^g})}$$
$$= \frac{\varrho_g(L; P_1) + \varrho_g(L; P_1') + \varrho_g(L; P_1'')}{3}, \tag{5.38}$$
$$\chi^{g^2}(\mathbb{P}^2; L) = \frac{\varrho_{g^2}(L; P_1) + \varrho_{g^2}(L; P_1') + \varrho_{g^2}(L; P_1'')}{3}.$$

It remains to compute the eigenvalues $\varrho_g(L; P)$ that originate in the equivariant Chern character $\mathrm{ch}_g(\mathcal{O}(k))$ of the line bundle, i.e. we need to determine the irreducible representation of $\mathcal{O}(k)$ under the $\mathbb{Z}_3$-action. Using the so-called process of homogenization (see section 5.4 of [46]) the divisor $\mathcal{D} = kH$ that defines the bundle $\mathcal{O}(\mathcal{D})$ can be represented by a monomial

$$Q_{\mathbb{P}^2}(kH) = x_1^a x_2^b x_3^{k-a-b}, \tag{5.39}$$

where $a, b, k \in \mathbb{Z}$. Whereas the strict definition of those monomials utilizes an inner product between certain lattice points related to the fan of $\mathbb{P}^2$ and the lattice points of the divisor $\mathcal{D}$, the above form of such monomials can be easily read of from the GLSM charges, see (5.50) and (5.56) in the later examples. One can interpret the space of global sections of $\mathcal{O}(kH)$ as generated by monomials of the form $x_1^a x_2^b x_3^c$ where $a + b + c = k$, i.e. we can effectively use the monomial as a representation of the bundle. This representation bears a striking resemblance to our rationoms, cf. (5.17). The idea is then to apply the different (equivalent) base involutions $g_1, g_1', g_1''$ associated to the fixpoints $P_1, P_1', P_1''$ on this monomial and determine the value picked up relative to the involution that we choose for the equivariant structure, i.e. the involution $g_1$ in this example. The choice of the equivariant structure for the bundle is therefore reflected in the bundle representation eigenvalues $\varrho_g(L; P)$. For our example we therefore have

$$Q = x_1^a x_2^b x_3^{k-a-b} \begin{array}{c} g_1 \\ \longrightarrow \\ \end{array} \overbrace{\alpha^{a+2b} Q}^{\text{value of equivariant structure}} \rightsquigarrow \varrho_g(\mathcal{O}(k); P_1) = 1 \tag{5.40}$$

$$\begin{array}{c} g_1' \\ \longrightarrow \end{array} \alpha^{-k} \alpha^{a+2b} Q \quad \rightsquigarrow \quad \varrho_g(\mathcal{O}(k); P_1') = \alpha^{-k}$$

$$\begin{array}{c} g_1'' \\ \longrightarrow \end{array} \alpha^{k} \alpha^{a+2b} Q \quad \rightsquigarrow \quad \varrho_g(\mathcal{O}(k); P_1'') = \alpha^{k}$$

and an analogous result for $g_2$, leading to the final expressions

$$\chi^g(\mathbb{P}^2; \mathcal{O}(k)) = \chi^{g^2}(\mathbb{P}^2; \mathcal{O}(k)) = \frac{1 + \alpha^k + \alpha^{-k}}{3} = \begin{cases} 1 & k \in 3\mathbb{Z} \\ 0 & \text{otherwise} \end{cases}. \tag{5.41}$$

Together with the ordinary Euler characteristic of $\mathcal{O}(k)$ on $\mathbb{P}^2$

$$\chi(\mathbb{P}^2; \mathcal{O}(k)) = 1 + \frac{k(k+3)}{2} \tag{5.42}$$

## 5.5. Some explicit examples for finite group equivariance

we therefore obtain the orientifold Euler characteristic

$$\chi(\mathbb{P}^2/\mathbb{Z}_3; \mathcal{O}(k)) = \frac{\chi + \chi^g + \chi^{g^2}}{3}$$
$$= \frac{6 + 3k(k+3) + 4(1 + \alpha^k + \alpha^{-k})}{18}, \quad (5.43)$$

which completes the computation on the well-established and proven mathematical side.

The idea is now to simply apply the involution mapping to the LMs of our counting algorithm and count the remaining invariant LMs. Recall from section 5.3 that this already implies a choice of the equivariant $\mathbb{Z}_3$-structure on the bundle, where we will use the non-primed involution mapping $g_1$. Consider for example the bundle $\mathcal{O}(-6)$ on $\mathbb{P}^2$. From (5.43) we expect to find $\chi(\mathbb{P}^2/\mathbb{Z}_3; \mathcal{O}(-6)) = 4$. The relevant LMs and their respective phases picked up from the involution are

$$\underbrace{\frac{1}{x_1^4 x_2 x_3}}_{g_1 \to 1}, \underbrace{\frac{1}{x_1 x_2^4 x_3}}_{g_1 \to 1}, \underbrace{\frac{1}{x_1 x_2 x_3^4}}_{g_1 \to 1}, \underbrace{\frac{1}{x_1^3 x_2^2 x_3}}_{g_1 \to \alpha}, \underbrace{\frac{1}{x_1^3 x_2 x_3^2}}_{g_1 \to \alpha^2},$$
$$\underbrace{\frac{1}{x_1^2 x_2^3 x_3}}_{g_1 \to \alpha^2}, \underbrace{\frac{1}{x_1 x_2^3 x_3^2}}_{g_1 \to \alpha}, \underbrace{\frac{1}{x_1^2 x_2 x_3^3}}_{g_1 \to \alpha}, \underbrace{\frac{1}{x_1 x_2^2 x_3^3}}_{g_1 \to \alpha^2}, \underbrace{\frac{1}{x_1^2 x_2^2 x_3^2}}_{g_1 \to 1},$$

$$h^2(\mathbb{P}^2; \mathcal{O}(-6)) = (4_{\text{inv}}, 3_\alpha, 3_{\alpha^2}) \quad (5.44)$$

yielding $h_{\text{inv}}^\bullet(\mathbb{P}^2; \mathcal{O}(-6)) = (0, 0, 4)$ and therefore the expected result for the Euler characteristic of the orbifold space $\mathbb{P}^2/\mathbb{Z}_3$. Note that due to $g_2 = g_1^2$ one only has to evaluate the effect of the generators to identify the invariant rationoms. This agreement has been checked for a wide range of bundles $\mathcal{O}(k)$ on $\mathbb{P}^2/\mathbb{Z}_3$.

**Example 5.5.2** (Example: $dP_1/\mathbb{Z}_3$). Next we consider a blowup of $\mathbb{P}^2$, i.e. the del Pezzo-1 surface. The involution (5.29) basically remains unchanged, acting now on the four homogeneous coordinates of $dP_1$ as

$$g : (x_1, x_2, x_3, x_4) \mapsto (\alpha x_1, \alpha^2 x_2, x_3, x_4) \quad \text{for } \alpha := \sqrt[3]{1} = e^{\frac{2\pi i}{3}}. \quad (5.45)$$

Following from the projective equivalences listed in table 5.2, we can identify the four fixpoints of the action:

$$P_1 = (1, 0, 0, 1), \quad P_2 = (0, 1, 0, 1), \quad P_3 = (0, 1, 1, 0), \quad P_4 = (1, 0, 1, 0). \quad (5.46)$$

By using local charts around those fixpoints like in (5.33), the tangent space map-

| vertices of the polyhedron / fan | coords | GLSM charges $Q^m$ $Q^n$ | | divisor class |
|---|---|---|---|---|
| $v_1 = (-1, -1)$ | $x_1$ | 1 | 0 | $H$ |
| $v_2 = (\phantom{-}1, \phantom{-}0)$ | $x_2$ | 1 | 0 | $H$ |
| $v_3 = (\phantom{-}0, \phantom{-}1)$ | $x_3$ | 1 | 1 | $H + X$ |
| $v_4 = (\phantom{-}0, -1)$ | $x_4$ | 0 | 1 | $X$ |

intersection form: $HX - X^2$

$$\text{SR}(dP_1) = \langle x_1 x_2, \, x_3 x_4 \rangle$$

**Table 5.2.:** Toric data for the del Pezzo-1 surface

ping eigenvalues reveal the following representations for the conjugated normal bundles:

$$\bar{N}_{P_1} = \bar{N}_{P_1}^\alpha \oplus \bar{N}_{P_1}^{\alpha^2}, \quad \bar{N}_{P_2} = \bar{N}_{P_2}^\alpha \oplus \bar{N}_{P_2}^{\alpha^2}, \quad \bar{N}_{P_3} = (\bar{N}_{P_3}^\alpha)^2, \quad \bar{N}_{P_4} = (\bar{N}_{P_4}^{\alpha^2})^2. \tag{5.47}$$

Compared to the three $\mathbb{P}^2$ fixpoints of the analogous $\mathbb{Z}_3$-action, whose representations were all of the type $\bar{N}_P^\alpha \oplus \bar{N}_P^{\alpha^2}$, the additional blowup of $dP_1$ seems to split up the contribution of one of three $\mathbb{P}^2$ fixpoints. This can be seen by

$$\begin{aligned} \text{ch}_g\big((\bar{N}_P^\alpha)^2\big)\big|_P &= 1 - \alpha \cdot 2 + \alpha^2 = (1 - \alpha)^2 \\ \text{ch}_g\big((\bar{N}_P^{\alpha^2})^2\big)\big|_P &= 1 - \alpha^2 \cdot 2 + \alpha^4 = (1 - \alpha^2)^2 \end{aligned} \tag{5.48}$$

and noting that the sum of both these contributions adds up to

$$\frac{1}{(1-\alpha)^2} + \frac{1}{(1-\alpha^2)^2} = \frac{1}{3}, \tag{5.49}$$

i.e. precisely the contribution that each $\mathbb{P}^2$ fixpoint added to the Lefschetz numbers in the previous example. The global sections of the bundle $\mathcal{O}(m,n)$ over $dP_1$ can be represented by monomials of the form

$$Q_{dP_1}(mH + nX) = x_1^a x_2^b x_3^{m-a-b} x_4^{n-m+a+b}, \tag{5.50}$$

and relative to the involution $g$ from (5.45) (that we choose for the equivariant structure) this gives the relative signs, fixpoints and normal bundle representation

## 5.5. Some explicit examples for finite group equivariance 113

splittings

$$
\begin{aligned}
P_1 &= (1,0,0,1) & \bar{N}_{P_1} &= \bar{N}_{P_1}^\alpha \oplus \bar{N}_{P_1}^{\alpha^2} & \tfrac{1}{3} & & \alpha^{-m} \\
P_2 &= (0,1,0,1) & \bar{N}_{P_2} &= \bar{N}_{P_2}^\alpha \oplus \bar{N}_{P_2}^{\alpha^2} & \tfrac{1}{3} & & \alpha^{m} \\
P_3 &= (0,1,1,0) & \bar{N}_{P_3} &= (\bar{N}_{P_3}^\alpha)^2 & \tfrac{1}{(1-\alpha)^2} & & \alpha^{m-n} \\
P_4 &= (1,0,1,0) & \bar{N}_{P_4} &= (\bar{N}_{P_4}^{\alpha^2})^2 & \tfrac{1}{(1-\alpha^2)^2} & & \alpha^{-(m-n)}
\end{aligned}
\tag{5.51}
$$

From the standard **Riemann-Roch-Hirzebruch** formula (5.1) one can compute the ordinary Euler characteristic

$$\chi(dP_1; \mathcal{O}(m,n)) = 1 + m + mn + \frac{1}{2}n(1-n) \tag{5.52}$$

and from the fixpoint data listed in (5.51) the Lefschetz number of the generator $g$ can be evaluated as

$$\chi^g(dP_1; \mathcal{O}(m,n)) = \frac{\alpha^{-m} + \alpha^m}{3} + \frac{\alpha^{m-n}}{(1-\alpha)^2} + \frac{\alpha^{-(m-n)}}{(1-\alpha^2)^2}. \tag{5.53}$$

Note that this Lefschetz number is not an integer for generic values of $m, n \in \mathbb{Z}$. However, since for the $\mathbb{Z}_3$ group the direct identification (5.7) of the single Lefschetz number $\chi^g$ with dimensions of cohomology groups is no longer given, this does not pose a problem. One can show that the Lefschetz number for the second non-unit group element $g^2 \in \mathbb{Z}_3$ can be obtained from replacing $\alpha \to \alpha^2$ in formula (5.53). Ultimately, we therefore arrive at the Euler characteristic

$$
\begin{aligned}
\chi(dP_1/\mathbb{Z}_3; \mathcal{O}(m,n)) &= \frac{\chi + \chi^g + \chi^{g^2}}{3} \\
&= \frac{1}{3}\left[1 + m + mn + \frac{n(1-n)}{2} + \frac{2(\alpha^{-m} + \alpha^m)}{3} + \frac{2\alpha^{m-n}}{(1-\alpha)^2} + \frac{2\alpha^{-(m-n)}}{(1-\alpha^2)^2}\right]
\end{aligned}
\tag{5.54}
$$

for the orbifold space obtained from the $\mathbb{Z}_3$-action on the single blowup of $\mathbb{P}^2$. Turning to the counting of $g$-invariant LMs from our algorithm as in (5.44), we find once again perfect agreement with the Euler characteristic derived from the Lefschetz theorem above.

**Example 5.5.3** (Example: $dP_3/\mathbb{Z}_3$). The natural extension to the previous example is to blowup the $dP_1$ twice further, giving us the $dP_3$ surface with the toric data in table 5.3. Once again we employ the same extension of the action (5.29) on the base. However, in order to simplify the subsequent computation, this time

| vertices of the polyhedron / fan | coords | GLSM charges $Q^m$ $Q^n$ $Q^p$ $Q^q$ | divisor class |
|---|---|---|---|
| $v_1 = (-1, \ -1)$ | $x_1$ | 1 0 0 1 | $H + Z$ |
| $v_2 = (\ \ 1, \ \ \ 0)$ | $x_2$ | 1 0 1 0 | $H + Y$ |
| $v_3 = (\ \ 0, \ \ \ 1)$ | $x_3$ | 1 1 0 0 | $H + X$ |
| $v_4 = (\ \ 0, \ -1)$ | $x_4$ | 0 1 0 0 | $X$ |
| $v_5 = (-1, \ \ \ 0)$ | $x_5$ | 0 0 1 0 | $Y$ |
| $v_6 = (\ \ 1, \ \ \ 1)$ | $x_6$ | 0 0 0 1 | $Z$ |

intersection form: $\quad HX + HY + HZ - 2H^2 - X^2 - Y^2 - Z^2$

$\text{SR}(dP_3) = \langle x_1 x_2, \ x_1 x_3, \ x_1 x_6, \ x_2 x_3, \ x_2 x_5, \ x_3 x_4, \ x_4 x_5, \ x_4 x_6, \ x_5 x_6 \rangle$

**Table 5.3.:** Toric data for the del Pezzo-3 surface.

we choose a different equivariant structure, which is induced via

$$g : (x_1, \ldots, x_6) \mapsto (x_1, \alpha x_2, x_3, \alpha x_4, x_5, x_6) \\ \sim (\alpha x_1, \alpha^2 x_3, x_3, x_4, x_5, x_6). \quad \text{for } \alpha := \sqrt[3]{1} = e^{\frac{2\pi i}{3}}. \quad (5.55)$$

The $\mathbb{Z}_3$-action on $dP_3$ reveals six fixpoints which can be related to the "splitting" of each of the three $\frac{1}{3}$-fixpoint contributions from the original $\mathbb{P}^2$ computation. Computing the induced normal bundle representation and relative signs via

$$Q_{dP_3}(mH + nX + pY + qZ) = x_1^a x_2^b x_3^{m-a-b} x_4^{n-m+a+b} x_5^{p-b} x_6^{q-a} \quad (5.56)$$

is completely analogous, albeit quite laborious, to the previous cases and yields the following fixpoint data:

$$\begin{aligned}
P_1 &= (1,0,1,0,1,1) & \bar{N}_{P_1} &= (\bar{N}^{\alpha^2}_{P_1})^2 & \tfrac{1}{(1-\alpha^2)^2} & & 1, \\
P_2 &= (0,1,1,1,0,1) & \bar{N}_{P_2} &= (\bar{N}^{\alpha^2}_{P_2})^2 & \tfrac{1}{(1-\alpha^2)^2} & & \alpha^{m-n+p}, \\
P_3 &= (1,1,0,1,1,0) & \bar{N}_{P_3} &= (\bar{N}^{\alpha^2}_{P_3})^2 & \tfrac{1}{(1-\alpha^2)^2} & & \alpha^{q-m-n}, \\
P_4 &= (0,1,1,0,1,1) & \bar{N}_{P_4} &= (\bar{N}^{\alpha}_{P_4})^2 & \tfrac{1}{(1-\alpha)^2} & & \alpha^{n-m}, \\
P_5 &= (1,0,1,1,1,0) & \bar{N}_{P_5} &= (\bar{N}^{\alpha}_{P_5})^2 & \tfrac{1}{(1-\alpha)^2} & & \alpha^{m-n-q}, \\
P_6 &= (1,1,0,1,0,1) & \bar{N}_{P_6} &= (\bar{N}^{\alpha}_{P_6})^2 & \tfrac{1}{(1-\alpha)^2} & & \alpha^{-n-p}.
\end{aligned} \quad (5.57)$$

Employing once again the well-known Riemann-Roch-Hirzebruch formula, the

## 5.5. Some explicit examples for finite group equivariance

Euler characteristic of $dP_3$ turns out to be

$$\chi(dP_3, \mathcal{O}(m,n,p,q)) = 1 - m^2 + mn + mp + mq \\ + \frac{n(1-n) + p(1-p) + q(1-q)}{2} \tag{5.58}$$

and from the fixpoint data in (5.58) we can compute the Lefschetz number

$$\chi^g(dP_3; \mathcal{O}(m,n,p,q)) = \frac{1 + \alpha^{m-n+p} + \alpha^{q-m-n}}{(1-\alpha^2)^2} \\ + \frac{\alpha^{n-m} + \alpha^{m-n-q} + \alpha^{-n-p}}{(1-\alpha)^2}. \tag{5.59}$$

Again, this number will not be an integer for a generic choice of the bundle divisor $D = mH + nX + pY + qZ$. The second Lefschetz number $\chi^{g^2}$ can be obtained by replacing $\alpha \to \alpha^2$ in formula (5.59), such that the average of the three terms gives us the Euler characteristic of the orbifold space $dP_3/\mathbb{Z}_3$:

$$\chi(dP_3/\mathbb{Z}_3; \mathcal{O}(m,n,p,q)) = 1 - m^2 + mn + mp + mq \\ + \frac{n(1-n) + p(1-p) + q(1-q)}{2} + \frac{1}{3} \\ + \frac{\alpha^{m-n} + \alpha^{n+p} + \alpha^{m-n+p} + \alpha^{-m+n+q} + \alpha^{-m-n+q}}{(1-\alpha^2)^2} \\ + \frac{\alpha^{-m+n} + \alpha^{-n-p} + \alpha^{-m+n-p} + \alpha^{m-n-q} + \alpha^{m+n-q}}{(1-\alpha)^2}. \tag{5.60}$$

By comparison to the LM counting of our algorithm, we find once again perfect agreement. In addition to simply providing a more complicated example, the $dP_3$ LM counting also involves non-trivial multiplicity factors 2. Consider for example the line bundle $\mathcal{O}(-5,-1,-1,-1)$, which has six monomials each contributing with multiplicity factor 2. Applying the involution (5.55) (that was chosen for the equivariant structure and therefore directly acts on the LMs) we observe the following:

$$2 \times \left( \underbrace{\frac{x_4^2}{x_1 x_2 x_3^3}}_{g \to \alpha}, \underbrace{\frac{x_4 x_5}{x_1 x_2^2 x_3^2}}_{g \to \alpha^2}, \underbrace{\frac{x_5^2}{x_1 x_2^3 x_3}}_{g \to 1}, \underbrace{\frac{x_4 x_6}{x_1^2 x_2 x_3^2}}_{g \to 1}, \underbrace{\frac{x_5 x_6}{x_1^2 x_2^2 x_3}}_{g \to \alpha}, \underbrace{\frac{x_6^2}{x_1^3 x_2 x_3}}_{g \to \alpha^2} \right) \tag{5.61}$$

$$\underbrace{\phantom{xxxxxxxxxxxxxxxxxxxxxxxxxxxxxxxxxxxxxxxxxxxxxxxxxxxxxxxxxxx}}_{h^1(dP_3; \mathcal{O}(-5,-1,-1,-1)) = (4_{\text{inv}}, 4_\alpha, 4_{\alpha^2})}$$

Plugging the bundle charges into (5.60) yields

$$\chi(dP_3/\mathbb{Z}_3; \mathcal{O}(-5,-1,-1,-1)) = -4, \tag{5.62}$$

once again in agreement with the result obtained from the counting of invariant LMs. Similar to the observation made for $\mathbb{Z}_2$-equivariant situations we therefore find the same "canonical" behavior of such LMs, i.e. an invariant LM with multiplicity 2 simply contributes twice to the counting of invariant LMs.

## 5.6 Generalized equivariant algorithm conjecture

The steps involved in the computation of the Lefschetz character in the $\mathbb{P}^2$ example are completely analogous on $\mathbb{P}^{n-1}$ with the group $\mathbb{Z}_n$ and the base space generator involution

$$g : x_i \mapsto \alpha^i x_i \qquad \text{with } \alpha := \sqrt[p]{1} = e^{\frac{2\pi i}{p}}. \tag{5.63}$$

We have successfully checked this for various values of $n$ and bundles $\mathcal{O}(k)$. Together with the empirical evidence gathered from the presented and various other examples, we therefore arrive at the following hypothesis which generalizes the conjecture from section 5.6:

**Conjecture 5.6.1** (*G*-equivariant cohomology calculation of finite groups:). Let $\mathbb{P}_\Sigma$ be a toric variety and $\mathcal{O}_{\mathbb{P}_\Sigma}(\mathcal{D})$ some line bundle over $\mathbb{P}_\Sigma$. Consider furthermore the generator involutions on the base $\sigma_1, \ldots, \sigma_r : \mathcal{X} \longrightarrow \mathcal{X}$ which allow for an equivariant structure on the line bundle $\mathcal{O}_{\mathbb{P}_\Sigma}(\mathcal{D})$. The cohomology of $\mathcal{O}_{\mathbb{P}_\Sigma}(\mathcal{D})$ decomposes according to theorem 4.6.1 into a direct sum of spaces that arise according to its contributing Laurent monomials:

$$\text{Contributing LMs} = \{s_1, \ldots, s_N\}$$
$$\Rightarrow H^\bullet(\mathbb{P}_\Sigma; \mathcal{O}_{\mathbb{P}_\Sigma}(\mathcal{D})) = \bigoplus_{k=1}^{N} H^\bullet_{s_k}(\mathbb{P}_\Sigma; \mathcal{O}_{\mathbb{P}_\Sigma}(\mathcal{D})). \tag{5.64}$$

The equivariant cohomology groups are then given by the direct sum of contributing spaces that correspond to Laurent monomials that are themselves invariant under the generators of $G$ and those that are not, i.e. if we put the $s_k$ in (5.64) in order such that the first $\tilde{k}$ are invariant and the last $N - \tilde{k}$ are non-invariant after applying all generators of $G$ on the homogeneous coordinates in the LMs,

## 5.6. Generalized equivariant algorithm conjecture

we find

$$\begin{aligned}
H^\bullet(\mathbb{P}_\Sigma; \mathcal{O}_{\mathbb{P}_\Sigma}(\mathcal{D})) &= H^\bullet_{\text{inv}}(\mathbb{P}_\Sigma; \mathcal{O}_{\mathbb{P}_\Sigma}(\mathcal{D})) \oplus H^\bullet_{\text{non-inv}}(\mathbb{P}_\Sigma; \mathcal{O}_{\mathbb{P}_\Sigma}(\mathcal{D})), \\
H^\bullet(\mathbb{P}_\Sigma/G; \mathcal{O}_{\mathbb{P}_\Sigma/G}(\mathcal{D})) &= H^\bullet_{\text{inv}}(\mathbb{P}_\Sigma; \mathcal{O}_{\mathbb{P}_\Sigma}(\mathcal{D})) = \bigoplus_{k=1}^{\bar{k}} H^\bullet_{s_k}(\mathbb{P}_\Sigma; \mathcal{O}_{\mathbb{P}_\Sigma}(\mathcal{D})), \\
H^\bullet_{\text{non-inv}}(\mathbb{P}_\Sigma; \mathcal{O}_{\mathbb{P}_\Sigma}(\mathcal{D})) &= \bigoplus_{k=\bar{k}}^{N} H^\bullet_{s_k}(\mathbb{P}_\Sigma; \mathcal{O}_{\mathbb{P}_\Sigma}(\mathcal{D}))
\end{aligned} \quad (5.65)$$

At this point we would like to emphasize the tremendous computational power of this conjecture. Already for the $dP_3$ example which is still a rather simple surface, we see that the computations necessary to just determine the Euler characteristic (much less than the cohomology groups themselves) from the established mathematics is quite enduring, whereas this computation for reasonably low values of the bundle charges can be done via pen and paper in a couple of minutes. As before the efficiency and ease-of-usage this conjecture provides ultimately rests on the direct applicability of the involution mappings defined on the coordinates of the base on the LMs representing the cohomology.

# Chapter 6

# Subvarieties and Calabi-Yau Manifolds

Toric varieties are really convenient to handle as we have seen and we could write down many nice formulæ to calculate several of their topological quantities. Nevertheless, we will see in the remainder of this section that it is also quite important to carry our machinery over to subvarieties. Luckily we can usually get these from well-understood calculations that are performed on the toric variety itself. So in this section we will introduce all the methods necessary in order to obtain properties of the subvariety from those of the ambient toric variety.

## 6.1 Physical motivation: Calabi-Yau compactifications, D-branes and GUT divisors in F-theory

The motivation to introduce all the mathematics in this chapter lies in the construction of heterotic string models as we described in chapter 2. There we saw the need for a very specific kind of geometric space to obtain sensible compactifications of such theories. Such spaces are called *Calabi-Yau manifolds* and have a couple of equivalent definitions which we will introduce in the beginning of the next section. One can show that toric varieties are not suited as compactification spaces since they are either Calabi-Yau or compact but never both and we need both for the purpose of compactification. Hence we need to consider subvarieties of toric varieties. Not only in the heterotic setting but also in type II string theories subvarieties of toric spaces will arise naturally. For instance D-branes that live on a Calabi-Yau space can be described as subvarieties of the Calabi-Yau and hence as higher codimensional subvarieties of the toric variety in our setup. Another example is the GUT divisor and the matter curve in F-theory arising as codimension two and three subvarieties of a Calabi-Yau four-fold respectively.

## 6.2 Calabi-Yau spaces

**Definition 6.2.1** (Calabi-Yau). Let $\mathcal{M}$ be a complex $d_\mathcal{M}$-dimensional Kähler manifold. $\mathcal{M}$ is called *Calabi-Yau*, iff

$\mathcal{M}$ has vanishing first Chern class: $c_1(\mathcal{M}) := c_1(T_\mathcal{M}) = 0$.

$\overset{\text{Yau}}{\Leftrightarrow}$ The Kähler metric has $SU(3)$ holonomy and with respect to this metric $\mathcal{M}$ is Ricci flat, i.e. the Ricci tensor vanishes: $\text{Ric}(\mathcal{M}) = 0$.

$\Leftrightarrow$ $\mathcal{M}$ admits for a holomorphic nowhere vanishing covariantly constant spinor.

Here the equivalence between the first and the second definition is highly nontrivial and was conjectured by Calabi in 1957 [104] before it was proven twenty years later by Yau [82]. For heterotic string theory in particular the first of these equivalent statements will arise as condition on the so-called target space of the corresponding non-linear sigma model (more about this in chapter 2). Therefore this is the property of Calabi-Yau spaces we will be using throughout the remainder.

We have seen a lot about explicit calculations on toric varieties and we want to have a Calabi-Yau space for string compactifications. Hence it would be a great thing to use toric varieties that have the Calabi-Yau property. But there is a problem here which we state in following theorem.

**Theorem 6.2.1.** *A toric variety can only be Calabi-Yau if it is non-compact.*

So we have actually no chance to build compact toric varieties that have the Calabi-Yau property. But in order to compactify the space time in which the heterotic string lives, we need such a Calabi-Yau to be compact. Are chapters 3 and 4 actually useless for string theorists? Well, luckily they are not. Although it is not possible to realize a compact Calabi-Yau manifold as a toric variety, it is still possible to realize it as a subvariety of a toric variety. More specifically one can obtain it for instance as a hypersurface inside a toric variety by imposing a constraint in form of an equation. One can also introduce more than one constraint and if one does it in a way such that the corresponding hypersurfaces intersect each other orthogonally then one obtains a proper subvariety with a unique dimension. Such subvarieties are then called complete intersections. In terms of homogeneous degrees of these hypersurfaces on can reformulate the Calabi-Yau condition in the following way:

**Theorem 6.2.2.** *Let $\mathbb{P}_\Sigma$ be a toric variety with coordinates $x_1, ..., x_n$ with degrees $Q_1{}^\alpha, ..., Q_n{}^\alpha$. Let furthermore $G_1, ..., G_c$ be homogeneous polynomials in $\mathbb{P}_\Sigma$. The complete intersection $\mathcal{S}$ of the corresponding subvarieties*

$$\mathcal{S} = \{G_1 = 0\} \cap ... \cap \{G_c = 0\} \tag{6.1}$$

*has vanishing first Chern class if and only if the homogeneous degrees of the hypersurfaces sum up to sum of the homogeneous degrees of the coordinates of $\mathbb{P}_\Sigma$, i.e.*

$$\mathcal{S} \text{ is Calabi-Yau} \Leftrightarrow ||G_1||^\alpha + ... + ||G_c||^\alpha = Q_1{}^\alpha + ... + Q_n{}^\alpha \quad \forall \alpha. \tag{6.2}$$

**Example 6.2.1.** The easiest and most studied example of such a Calabi-Yau manifold is given for the simple projective space $\mathbb{P}^4$. Here for five coordinate we have $Q_1 = 1$. Hence in order to define a hypersurface that is Calabi-Yau we need a polynomial of homogeneous degree five and for instance

$$G = x_1^5 + x_2^5 + x_3^5 + x_4^5 + x_5^5 \tag{6.3}$$

already does the trick[1]. Since we have a polynomial of degree five, the space $\mathcal{M} := \{G = 0\}$ is usually referred to as the Quintic.

In the following we will describe how we can derive the topological quantities introduced in the chapters 3 and 4 now for subvarieties that are given by the intersection of the vanishing sets of any homogeneous polynomials. Here we will not make use of the Calabi-Yau property and the methods will apply generally.

## 6.3 Line bundles and their cohomology

In chapter 4 we have introduced our algorithm that allows to calculate the cohomology of line bundles in a very efficient way. Since we have control over this part we would like to see how we can get the cohomology of arbitrary line bundles that live on a subvariety. In fact all the information about it is encoded in the cohomology of line bundles over the ambient toric variety and hence these are roughly all the ingredients one needs.

---

[1] You may of course choose a different polynomial. The only restriction here is that is gives rise to a smooth subvariety.

### 6.3.1 The Koszul sequence for hypersurfaces

Consider an irreducible hypersurface $\mathcal{S} \subset \mathbb{P}_\Sigma$ given by some homogeneous polynomial $G(x_1, ..., x_n)$, i.e.

$$\mathcal{S} := \{(x_1, ..., x_n) \in \mathbb{P}_\Sigma \mid G(x_1, ..., x_n) = 0\} \; . \tag{6.4}$$

For the remainder we will use $\mathcal{S}$ as a place holder for the actual subvariety as in (6.4) but at the same time as a place holder for the homogeneous degree of the corresponding polynomial, i.e. we also define $\mathcal{S} := ||G||$. It should be clear from the context when we are referring to the homogeneous degree and when to the actual subvariety.

To see how we can relate the cohomology of a line bundle on such a subvariety $\mathcal{S}$ to the one on the ambient space $\mathbb{P}_\Sigma$, we have to make use of the fact that the divisor that corresponds to the codimension one subvariety $\mathcal{S}$ gives also rise to a line bundle $\mathcal{O}_{\mathbb{P}_\Sigma}(\mathcal{S})$, as explained in 3.4. There we saw that global sections $s$ of such a line bundle are elements in the zeroth cohomology group

$$s \in H^0(\mathbb{P}_\Sigma; \mathcal{O}(\mathcal{S})) \tag{6.5}$$

which are simply given by monomials in the homogeneous coordinates of $\mathbb{P}_\Sigma$ that have a multi-degree corresponding to the multi-degree of defining equation $\{G = 0\}$ of the divisor $\mathcal{S}$, i.e.

$$||G|| = ||s|| \; . \tag{6.6}$$

Similarly, the elements of higher cohomology groups are given by Laurent-monomials with the right degree. Hence $G$ can be considered as a map from one line bundle to a different one, which in the simplest case reads

$$\mathcal{O}_{\mathbb{P}_\Sigma} \xrightarrow{G} \mathcal{O}_{\mathbb{P}_\Sigma}(\mathcal{S}) \; , \tag{6.7}$$

mapping the structure sheaf $\mathcal{O}_{\mathbb{P}_\Sigma}$ of $\mathbb{P}_\Sigma$ to the line bundle that corresponds to $\mathcal{S}$. One can also perform the dual version of this and map the dual line bundle of $\mathcal{O}_{\mathbb{P}_\Sigma}(\mathcal{S})$ to the structure sheaf:

$$\mathcal{O}_{\mathbb{P}_\Sigma}(-\mathcal{S}) \xrightarrow{G} \mathcal{O}_{\mathbb{P}_\Sigma} \; . \tag{6.8}$$

The nice thing, already indicated by the arrow, is now that this is actually an injective map. Furthermore the restriction map $R$ restricts from the structure sheaf of $\mathbb{P}_\Sigma$ to the structure sheaf $\mathcal{O}_\mathcal{S} \subset \mathcal{O}_{\mathbb{P}_\Sigma}$ of $\mathcal{S}$ and it is therefore surjective.

## 6.3. Line bundles and their cohomology

So we have built a sequence of line bundles that is short and exact:

$$0 \longrightarrow \mathcal{O}_{\mathbb{P}_\Sigma}(-\mathcal{S}) \xrightarrow{G} \mathcal{O}_{\mathbb{P}_\Sigma} \xrightarrow{R} \mathcal{O}_\mathcal{S} \longrightarrow 0. \quad (6.9)$$

That it is exact, i.e. $G \circ R = 0$ is clear since we first map an element of $\mathcal{O}_{\mathbb{P}_\Sigma}(-\mathcal{S})$ basically by multiplication of $G$ to an element of $\mathcal{O}_{\mathbb{P}_\Sigma}$ and then put $G$ to zero with the restriction map $R$:

$$(G \circ R)(s) = R(G \cdot s) = G \cdot s|_{G=0} = 0, \text{ for some section } s \text{ of } \mathcal{O}_{\mathbb{P}_\Sigma}(-\mathcal{S}). \quad (6.10)$$

In order to relate an arbitrary line bundle on the hyper surface, say $\mathcal{O}_\mathcal{S}(\mathcal{D})$ with line bundles on the ambient space we simply tensor the sequence (6.9) with it:

$$0 \longrightarrow \mathcal{O}_{\mathbb{P}_\Sigma}(-\mathcal{S}) \otimes \mathcal{O}_{\mathbb{P}_\Sigma}(\mathcal{D}) \xrightarrow{G} \mathcal{O}_{\mathbb{P}_\Sigma} \otimes \mathcal{O}_{\mathbb{P}_\Sigma}(\mathcal{D}) \xrightarrow{R} \mathcal{O}_\mathcal{S} \otimes \mathcal{O}_{\mathbb{P}_\Sigma}(\mathcal{D}) \longrightarrow 0 \quad (6.11)$$

which gives us the following short exact sequence:

$$0 \longrightarrow \mathcal{O}_{\mathbb{P}_\Sigma}(-\mathcal{S}+\mathcal{D}) \xrightarrow{G} \mathcal{O}_{\mathbb{P}_\Sigma}(\mathcal{D}) \xrightarrow{R} \mathcal{O}_\mathcal{S}(\mathcal{D}) \longrightarrow 0. \quad (6.12)$$

This is called the *Koszul sequence* for an arbitrary line bundle on the hyper surface $\mathcal{S}$. Since it is exact and also short, due to theorem 4.3.2 there also exists a corresponding long exact sequence in cohomology:

$$\begin{aligned}
0 &\longrightarrow H^0(\mathbb{P}_\Sigma; \mathcal{O}_{\mathbb{P}_\Sigma}(-\mathcal{S}+\mathcal{D})) \longrightarrow H^0(\mathbb{P}_\Sigma; \mathcal{O}_{\mathbb{P}_\Sigma}(\mathcal{D})) \longrightarrow H^0(\mathcal{D}; \mathcal{O}_\mathcal{S}(\mathcal{D})) \\
&\longrightarrow H^1(\mathbb{P}_\Sigma; \mathcal{O}_{\mathbb{P}_\Sigma}(-\mathcal{S}+\mathcal{D})) \longrightarrow H^1(\mathbb{P}_\Sigma; \mathcal{O}_{\mathbb{P}_\Sigma}(\mathcal{D})) \longrightarrow H^1(\mathcal{D}; \mathcal{O}_\mathcal{S}(\mathcal{D})) \\
&\longrightarrow H^2(\mathbb{P}_\Sigma; \mathcal{O}_{\mathbb{P}_\Sigma}(-\mathcal{S}+\mathcal{D})) \longrightarrow H^2(\mathbb{P}_\Sigma; \mathcal{O}_{\mathbb{P}_\Sigma}(\mathcal{D})) \longrightarrow H^2(\mathcal{D}; \mathcal{O}_\mathcal{S}(\mathcal{D})) \longrightarrow \ldots
\end{aligned} \quad (6.13)$$

The maps $G$ and $R$ in the short sequence are of course essential for the mappings in this sequence and in particular they induce mappings between the first and second column as well as the second and the third column respectively, i.e.

$$H^i(\mathbb{P}_\Sigma; \mathcal{O}_{\mathbb{P}_\Sigma}(\mathcal{S}+\mathcal{D})) \xrightarrow{G^i} H^i(\mathbb{P}_\Sigma; \mathcal{O}_{\mathbb{P}_\Sigma}(\mathcal{D})) \xrightarrow{R^i} H^i(\mathcal{S}; \mathcal{O}_\mathcal{S}(\mathcal{D})). \quad (6.14)$$

This is in fact the easier part if one wants to do the honest computation of the cohomology groups and it is a little more involved to figure out the coboundary maps [2]

$$H^i(\mathcal{D}; \mathcal{O}_\mathcal{D}) \xrightarrow{\delta^i} H^{i+1}(\mathbb{P}_\Sigma; \mathcal{O}_{\mathbb{P}_\Sigma}(-\mathcal{S}+\mathcal{D})). \quad (6.15)$$

---
[2] See [105] for a full mathematical account on the Koszul complex, in particular the mappings.

124                                  6. Subvarieties and Calabi-Yau Manifolds

In practice in order to calculate the cohomology groups $H^i(\mathcal{S}; \mathcal{O}_\mathcal{S}(\mathcal{D}))$ one would need to compute the image of $R^i$ which is equal to the complement of the image of $G^i$ in $H^i(\mathbb{P}_\Sigma; \mathcal{O}_{\mathbb{P}_\Sigma}(\mathcal{D}))$ and then add a space that is isomorphic to the image of $\delta$ in $H^{i+1}(\mathbb{P}_\Sigma; \mathcal{O}_{\mathbb{P}_\Sigma}(-\mathcal{S} + \mathcal{D}))$ which is itself equal to the kernel of $G^{i+1}$ by exactness:

$$H^i(\mathcal{S}; \mathcal{O}_\mathcal{S}(\mathcal{D})) \cong \operatorname{im}(R^i) \oplus \operatorname{im}(\delta^i) = \operatorname{im}(R^i) \oplus \ker(G^{i+1}) \tag{6.16}$$

**Example 6.3.1.** Let me at this point state a couple of examples to make the ideas more assessable. We begin with a rather trivial one which still gives an idea how it works. Choose the ambient space as $\mathbb{P}^2$ as we did in subsection 4.7.2 and furthermore choose the hypersurface from a homogeneous degree one polynomial:

$$\mathcal{S} = \left\{(x_1, x_2, x_3) \in \mathbb{P}^2 \mid G = x_1 + x_2 + x_3 = 0\right\}. \tag{6.17}$$

We will now show how to calculate the cohomology of the line bundle $\mathcal{O}_\mathcal{S}(3)$. In order to do that we have to employ the Koszul sequence (6.31) which we find in this case to have the form

$$0 \longrightarrow \mathcal{O}_{\mathbb{P}^2}(2) \xrightarrow{G} \mathcal{O}_{\mathbb{P}^2}(3) \xrightarrow{R} \mathcal{O}_\mathcal{S}(3) \longrightarrow 0, \tag{6.18}$$

which gives the corresponding long exact sequence in cohomology (6.13) and looks as

$$\cdots \longrightarrow H^i(\mathbb{P}^2; \mathcal{O}_{\mathbb{P}^2}(2)) \xrightarrow{G} H^i(\mathbb{P}^2; \mathcal{O}_{\mathbb{P}^2}(3)) \xrightarrow{R} H^i(\mathcal{S}; \mathcal{O}_\mathcal{S}(3)) \longrightarrow \cdots, \tag{6.19}$$

where the induced maps $G$ and $R$ between the $i^{\text{th}}$ cohomology groups are then just given by

$$G(s_2^i) = (x_1 + x_2 + x_3) \cdot s_2^i, \qquad s_2^i \in H^i_{\mathbb{P}^2}(2), \tag{6.20}$$

$$R(s_3^i) = \begin{cases} 0, \text{ if } s_3^i = (x_1 + x_2 + x_3) \cdot s_2^i \\ s_3^i, \text{ otherwise} \end{cases}, \qquad s_3^i \in H^i_{\mathbb{P}^2}(3). \tag{6.21}$$

Here one can see explicitly how the image of $G$ gets mapped to the kernel of $R$. Having in mind the computation of line bundle cohomology of $\mathbb{P}^2$ in example 4.7.1 we know that actually $H^i_{\mathbb{P}^2}(a)$ is zero for all $a, i > 0$ and we only get contributions

## 6.3. Line bundles and their cohomology

for $i = 0$ and hence these vector spaces are just generated by monomials,

$$H^0_{\mathbb{P}^2}(2) = \langle x_1^2,\ x_1x_2,\ x_2^2,\ x_1x_3,\ x_2x_3,\ x_3^2 \rangle, \tag{6.22}$$

$$H^0_{\mathbb{P}^2}(3) = \langle x_1^3,\ x_1^2x_2,\ x_1x_2^2,\ x_2^3,\ x_1^2x_3,\ x_1x_3^2,\ x_3^3,\ x_2^2x_3,\ x_2x_3^2,\ x_1x_2x_3 \rangle$$
$$=: \langle e_1,\ e_2,\ e_3,\ e_4,\ e_5,\ e_6,\ e_7,\ e_8,\ e_9,\ e_{10} \rangle \tag{6.23}$$

and have therefore dimension 6 and 10 respectively. Because of the vanishing of all higher-dimensional cohomologies, it is pretty easy to obtain the desired cohomology group since it is given by the image of $R$ which is isomorphic to the quotient of $H^0_{\mathbb{P}^2}(\mathcal{O}_S(3))$ by the image of $G$:

$$H^0_{\mathcal{S}}(\mathcal{O}_S(3)) = \mathrm{im}(R) = H^0_{\mathbb{P}^2}(3)|_{G=x_1+x_2+x_3=0} \tag{6.24}$$

$$= \langle e_1,\ e_2,\ e_3,\ e_4,\ e_5,\ e_6,\ e_7,\ e_8,\ e_9,\ e_{10} \rangle|_{G=0} \tag{6.25}$$

$$= \frac{\langle e_1,\ e_2,\ e_3,\ e_4,\ e_5,\ e_6,\ e_7,\ e_8,\ e_9,\ e_{10} \rangle}{\langle e_{1,2,5},\ e_{2,3,10},\ e_{3,4,8},\ e_{5,6,10},\ e_{8,9,10},\ e_{6,7,9} \rangle}, \tag{6.26}$$

where we use the notation $e_{i,j,k} := e_i + e_j + e_k$ to denote the image vectors of the base vectors of $H^0_{\mathbb{P}^2}(2)$ under $G$, for instance

$$G(x_1^2) = x^2 \cdot (x_1 + x_2 + x_3) = e_1 + e_2 + e_5 = e_{1,2,5}. \tag{6.27}$$

So the task of deriving $H^0_{\mathcal{S}}(\mathcal{O}_S(3))$ has basically reduced to the task of taking a quotient vector space and hence to find a set of vectors that complete the image of $G$ to form a base of $H^0_{\mathbb{P}^2}(\mathcal{O}_S(3))$. In our example here the first four base vectors already do the job and hence we obtain

$$\boxed{H^0_{\mathcal{S}}(\mathcal{O}_S(3)) = \langle e_1,\ e_2,\ e_3,\ e_4 \rangle = \langle x_1^3,\ x_1^2x_2,\ x_1x_2^2,\ x_2^3 \rangle .} \tag{6.28}$$

*Remark.* In this example we showed very detailed how to do the calculation properly. In practice one will not always need the actual base vectors of the desired cohomology class but one may simply be interested in the dimension of it. If so, one will not always be faced with the task of deriving all the maps and kernels explicitly but may use the exactness of the sequence to predict the dimension. In Example 6.3.1 for instance if we were only interested in the dimension we could have used the that the dimension of a quotient space $\dim A/B$ is simply $\dim A - \dim B$ which would have given us $h^0_{\mathcal{S}}(\mathcal{O}_S(3)) = 4$ without calculating the image of $G$.

## 6.3.2 The Koszul sequence for complete intersections

Everything we explained in 6.3.1 can now also be extended to the situation where subvariety $\mathcal{S}$, we consider, is not only a hupersurface but a complete intersection of hypersurfaces:

$$\mathcal{S} = \bigcap_{j=1}^{c} \mathcal{S}_j, \quad \text{where } \mathcal{S}_j := \{(x_1, ..., x_n) \in \mathbb{P}_\Sigma \mid G_j(x_1, ..., x_n) = 0\} \,. \tag{6.29}$$

This means that we have $c$ hypersurfaces that intersect transversely inside the toric ambient space $\mathbb{P}_\Sigma$. Therefore the dimension $d_S$ of $\mathcal{S}$ will be the dimension $d$ of $\mathbb{P}_\Sigma$ minus the codimension $c$ of $\mathcal{S}$. The generalization of the plain Koszul sequence (6.9) is then derived by successively application of (6.9) for each hypersurface $\mathcal{S}_i$. As we had for codimension one a sequence of 2+1 sheaves we will now have one additional sheaf for each additional codimension. Doing this procedure carefully one finds

$$0 \longrightarrow \mathcal{O}_{\mathbb{P}_\Sigma}\left(-\sum_{j=1}^{c} \mathcal{S}_j\right) \longrightarrow \bigoplus_{j_1<...<j_{c-1}} \mathcal{O}_{\mathbb{P}_\Sigma}\left(-\sum_{k=1}^{c-1} \mathcal{S}_{j_k}\right) \longrightarrow$$

$$\longrightarrow \bigoplus_{j_1<...<j_{c-2}} \mathcal{O}_{\mathbb{P}_\Sigma}\left(-\sum_{k=1}^{c-2} \mathcal{S}_{j_k}\right) \longrightarrow ... \longrightarrow \bigoplus_{j_1<j_2} \mathcal{O}_{\mathbb{P}_\Sigma}(-\mathcal{S}_{j_1} - \mathcal{S}_{j_2}) \longrightarrow \tag{6.30}$$

$$\longrightarrow \bigoplus_{j} \mathcal{O}_{\mathbb{P}_\Sigma}(-\mathcal{S}_j) \longrightarrow \mathcal{O}_{\mathbb{P}_\Sigma} \longrightarrow \mathcal{O}_\mathcal{S} \longrightarrow 0$$

Note that given $\Lambda^1 \bigoplus_{i_1} \mathcal{O}_{\mathbb{P}_\Sigma}(-\mathcal{S}_{i_1}) = \bigoplus_{i_1} \mathcal{O}_{\mathbb{P}_\Sigma}(-\mathcal{S}_{i_1})$ all line bundles prior in the sequence chain can be interpreted as higher exterior powers. Twisting the whole sequence by $\mathcal{O}_{\mathbb{P}_\Sigma}(\mathcal{D})$ leads to

$$0 \longrightarrow \mathcal{O}_{\mathbb{P}_\Sigma}\left(-\sum_{j=1}^{c} \mathcal{S}_j + \mathcal{D}\right) \longrightarrow ... \longrightarrow \bigoplus_{j_1<j_2} \mathcal{O}_{\mathbb{P}_\Sigma}(-\mathcal{S}_{j_1} - \mathcal{S}_{j_2} + \mathcal{D}) \longrightarrow$$

$$\longrightarrow \bigoplus_{j_1} \mathcal{O}_{\mathbb{P}_\Sigma}(-\mathcal{S}_{j_1} + \mathcal{D}) \longrightarrow \mathcal{O}_{\mathbb{P}_\Sigma}(\mathcal{D}) \longrightarrow \mathcal{O}_\mathcal{S}(\mathcal{D}) \longrightarrow 0 \,.$$

$$\tag{6.31}$$

The maps between the sheaves are basically the $G_j$ that have to be applied to each direct summand properly. The last map is as before the restriction to the subvariety. In contrast to the situation with a simple hypersurface, we are not finished yet, since the sequence (6.31) is not a short exact one and hence does not give rise to a long exact sequence in cohomology. But one can easily see that

## 6.3. Line bundles and their cohomology

an exact sequence of length $c+2$ yields $c$ short exact sequences:

$$0 \longrightarrow A_1 \xrightarrow{g_1} A_2 \xrightarrow{g_2} A_3 \xrightarrow{g_3} A_4 \xrightarrow{g_4} \cdots \xrightarrow{g_{c-1}} A_c \xrightarrow{g_c} A_{c+1} \xrightarrow{g_{c+1}} A_{c+2} \longrightarrow 0 \tag{6.32}$$

can be decomposed to

$$0 \longrightarrow A_1 \xrightarrow{g_1} A_2 \xrightarrow{g_2} \mathrm{im}(g_2) \longrightarrow 0, \tag{6.33}$$

$$0 \longrightarrow \mathrm{im}(g_2) \xhookrightarrow{\iota} A_3 \xrightarrow{g_3} \mathrm{im}(g_3) \longrightarrow 0, \tag{6.34}$$

$$\vdots \tag{6.35}$$

$$0 \longrightarrow \mathrm{im}(g_{c-1}) \xhookrightarrow{\iota} A_{c+1} \xrightarrow{g_{c+1}} A_{c+2} \longrightarrow 0, \tag{6.36}$$

where $\iota$ denotes the inclusion map. For the case of the Koszul sequence by using several auxiliary sheaves $\mathcal{I}_k$ which represent the image of the corresponding map, we find the following $c$ short exact sequences:

$$
\begin{aligned}
0 \longrightarrow \mathcal{O}_{\mathbb{P}_\Sigma}\left(-\sum_{j=1}^{c} \mathcal{S}_j + \mathcal{D}\right) &\hookrightarrow \bigoplus_{j_1<\ldots<j_{c-1}} \mathcal{O}_{\mathbb{P}_\Sigma}\left(-\sum_{k=1}^{c-1} \mathcal{S}_{j_k} + \mathcal{D}\right) \twoheadrightarrow \mathcal{I}_1 \longrightarrow 0, \\
0 \longrightarrow \mathcal{I}_1 &\hookrightarrow \bigoplus_{j_1<\ldots<j_{c-2}} \mathcal{O}_{\mathbb{P}_\Sigma}\left(-\sum_{k=1}^{c-2} \mathcal{S}_{j_k} + \mathcal{D}\right) \twoheadrightarrow \mathcal{I}_2 \longrightarrow 0, \\
&\vdots \\
0 \longrightarrow \mathcal{I}_{c-2} &\hookrightarrow \bigoplus_{j} \mathcal{O}_{\mathbb{P}_\Sigma}\left(-\mathcal{S}_i + \mathcal{D}\right) \twoheadrightarrow \mathcal{I}_{c-1} \longrightarrow 0, \\
0 \longrightarrow \mathcal{I}_{c-1} &\hookrightarrow \mathcal{O}_{\mathbb{P}_\Sigma}(\mathcal{D}) \twoheadrightarrow \mathcal{O}_{\mathcal{S}}(\mathcal{D}) \longrightarrow 0.
\end{aligned}
\tag{6.37}
$$

These are the ones we are actually going to use in explicit calculations. This means that in order to derive the dimensions of the cohomology groups of $\mathcal{O}_{\mathcal{S}}(\mathcal{D})$ we first have to write down all the required long exact sequences and derive the cohomologies of $c-1$ auxiliary sheaves.

**Example 6.3.2.** For instance, for a complete intersection $\mathcal{S}$ of two hypersurfaces $\mathcal{S}_1$ and $\mathcal{S}_2$, the splitting of the generalized Koszul sequence (6.37) is given by

$$0 \longrightarrow \mathcal{O}_{\mathbb{P}_\Sigma}(-\mathcal{S}_1 - \mathcal{S}_2 + \mathcal{D}) \xrightarrow{(G_1, G_2)} \mathcal{O}_{\mathbb{P}_\Sigma}(-\mathcal{S}_1 + \mathcal{D}) \oplus \mathcal{O}_{\mathbb{P}_\Sigma}(-\mathcal{S}_2 + \mathcal{D}) \xrightarrow{\binom{G_2}{G_1}} \mathcal{I}_1 \longrightarrow 0$$

$$0 \longrightarrow \mathcal{I}_1 \xhookrightarrow{\iota} \mathcal{O}_{\mathbb{P}_\Sigma}(\mathcal{D}) \xrightarrow{R} \mathcal{O}_{\mathcal{S}}(\mathcal{D}) \longrightarrow 0 \tag{6.38}$$

and hence calculated in two steps.

# Chapter 7

# Vector Bundle-Valued Cohomology

So far we have learned a lot about calculating the cohomology of line bundles for a large class of spaces. We have seen how to calculate them on toric varieties and subvarieties thereof and also how to obtain the equivariant cohomology of line bundles on toric varieties that allow for an additional discrete group action.

In this chapter we want to see how we can make use of these concepts in order to calculate the cohomology of higher-rank vector bundles. Starting with the tangent bundle of the toric variety will lead us to the methods to obtain also the tangent bundle cohomology of subvarieties. This is already enough to determine the full Hodge-diamond of a Calabi-Yau three-fold and will be extended to more generic holomorphic vector bundles that may even be easier to work with than the tangent bundle. One can construct such vector bundles conveniently via the so-called monad construction or by extension. Following the tangent bundle-valued cohomology we will also point out how to obtain the cohomology of vector bundles that are given by tensor products of the initial bundle. In this context $\Lambda^2 \mathcal{V}$ and $\mathcal{V} \otimes \mathcal{V}$ will be of particular interest.

## 7.1 Physical motivation: Heterotic GUTs, moduli spaces of heterotic and type IIA/B theories

In chapter 2 we saw that we need a Calabi-Yau three-fold for heterotic string compactifications and furthermore that another essential ingredient is a holomorphic vector bundle of rank three, four or five. Of course one can always choose the tangent bundle as a rank three bundle but one can also choose the vector bundle completely independent from the tangent bundle which will give more freedom on the model building side, see chapter 2 for more details. There we saw that from a rank three, four and five vector bundle we can obtain a GUT with gauge group $E_6$, $SO(10)$ and $SU(5)$ respectively and the latter two are phenomenologically more interesting than the $E_6$ case. In all cases the spectrum of our theory will crucially depend on the holomorphic vector bundle. But also for type II models that live on Calabi-Yaus we can employ the methods of this chapter since the number of moduli of such a theory is determined by the two independent Hodge numbers of the Calabi-Yau. These Hodge numbers are nothing but the dimension of the first and the second tangent bundle-valued cohomology groups of the Calabi -Yau.

## 7.2 Bundle-valued cohomology

In section 4.2 we have introduced the idea of de Rham cohomology for smooth manifolds as well as the concept of Dolbeault cohomology for complex manifolds. In fact there is a nice relation between Dolbeault cohomology and the plain Čech cohomology of holomorphic vector bundles. As we have seen in (4.18), the space of differential forms with respect to d is split into a direct sum of forms arising from the splitting of d into a holomorphic and an anti-holomorphic part $\partial$, $\bar{\partial}$:

$$H_{\text{dR}}^k(\mathcal{M}) = \bigoplus_{p+q=k} H^{p,q}(\mathcal{M}).$$

One can now show that the $(p,q)^{\text{th}}$ Dolbeault cohomology group is nothing else but the $q^{\text{th}}$ Čech cohomology group of the vector bundle of the $p^{\text{th}}$ exterior power of the cotangent bundle:

$$H^{p,q}(\mathcal{M}) = H^q(\mathcal{M}; \Omega_{\mathcal{M}}^{0,p}). \tag{7.1}$$

This statement is not constrained to exterior powers of the cotangent bundle but can be applied to arbitrary holomorphic vector bundles $\mathcal{V}$. This means that for anti-holomorphic differential forms that take values in the holomorphic vector bundle $\mathcal{V}$, we can extend the Dolbeault operator $\bar{\partial}$ to act on such forms and hence obtain a complex

$$0 \longrightarrow \mathcal{V} \xrightarrow{\bar{\partial}} \mathcal{V} \otimes \Omega^{0,1}(\mathcal{M}) \xrightarrow{\bar{\partial}} \mathcal{V} \otimes \Omega^{0,2}(\mathcal{M}) \xrightarrow{\bar{\partial}} \mathcal{V} \otimes \Omega^{0,3}(\mathcal{M}) \xrightarrow{\bar{\partial}} \cdots \tag{7.2}$$

whose cohomology equals the Čech cohomology of the holomorphic vector bundle, i.e.

$$H^q(\mathcal{M}; \mathcal{V}) = \frac{\ker(\bar{\partial})}{\text{im}(\bar{\partial})}, \text{ where} \tag{7.3}$$

$$\bar{\partial} : \mathcal{V} \otimes \Omega^{0,q}(\mathcal{M}) \longrightarrow \mathcal{V} \otimes \Omega^{0,q+1}(\mathcal{M}).$$

## 7.2. Bundle-valued cohomology

$$
\begin{array}{c}
H^0(\mathcal{M};\mathcal{O}) \\
H^0(\mathcal{M};T^*_\mathcal{M}) \quad H^1(\mathcal{M};\mathcal{O}) \\
H^0(\mathcal{M};\Lambda^2 T^*_\mathcal{M}) \quad H^1(\mathcal{M};T^*_\mathcal{M}) \quad H^2(\mathcal{M};\mathcal{O}) \\
H^0(\mathcal{M};\Lambda^3 T^*_\mathcal{M}) \quad H^1(\mathcal{M};\Lambda^2 T^*_\mathcal{M}) \quad H^2(\mathcal{M};T^*_\mathcal{M}) \quad H^3(\mathcal{M};\mathcal{O}) \quad (7.4)\\
H^1(\mathcal{M};\Lambda^3 T^*_\mathcal{M}) \quad H^2(\mathcal{M};\Lambda^2 T^*_\mathcal{M}) \quad H^3(\mathcal{M};T^*_\mathcal{M}) \\
H^2(\mathcal{M};\Lambda^3 T^*_\mathcal{M}) \quad H^3(\mathcal{M};\Lambda^2 T^*_\mathcal{M}) \\
H^3(\mathcal{M};\Lambda^3 T^*_\mathcal{M})
\end{array}
$$

**Table 7.1.:** General Hodge diamond of a three-fold

$$
\begin{array}{c}
H^0(\mathcal{M};\mathcal{O}) \\
H^1(\mathcal{M};\mathcal{O}) \quad H^1(\mathcal{M};\mathcal{O}) \\
H^2(\mathcal{M};\mathcal{O}) \quad H^1(\mathcal{M};T^*_\mathcal{M}) \quad H^2(\mathcal{M};\mathcal{O}) \\
H^3(\mathcal{M};\mathcal{O}) \quad H^2(\mathcal{M};T^*_\mathcal{M}) \quad H^2(\mathcal{M};T^*_\mathcal{M}) \quad H^3(\mathcal{M};\mathcal{O}) \\
H^2(\mathcal{M};\mathcal{O}) \quad H^1(\mathcal{M};T^*_\mathcal{M}) \quad H^2(\mathcal{M};\mathcal{O}) \\
H^1(\mathcal{M};\mathcal{O}) \quad H^1(\mathcal{M};\mathcal{O}) \\
H^0(\mathcal{M};\mathcal{O})
\end{array}
$$

**Table 7.2.:** Poincare duality applied to a general Hodge diamond of a three-fold

This way we can think of the $q^{\text{th}}$ Čech cohomology group of a bundle $\mathcal{V}$ simply as the $\mathcal{V}$-valued $\bar{\partial}$-closed anti-holomorphic differential forms that are at the same time not $\bar{\partial}$-exact.

**Example 7.2.1** (Hodge diamond of a three-fold). As a nice application we can now express the Hodge diamond (4.19) of a smooth manifold $\mathcal{M}$ simply in terms of the Čech cohomology of exterior powers of the cotangent bundle according to (7.1) which for a three-dimensional manifold reads as shown in table 7.1 which reduce due to complex conjugation and Poincare duality to the one shown in table 7.2. Therefore for a three-fold it is sufficient to calculate the Čech cohomology of the structure sheaf as well as the first and second cohomology groups of the tangent bundle. If we are dealing with a Calabi-Yau manifold we can simplify this even further and have to calculate only two independent cohomology groups as shown in table 7.3.

As we will see below the calculation of Čech cohomology of a holomorphic vector bundle can in many cases be reduced to the calculation of line bundle cohomology of various line bundles and therefore we can apply theorem 4.6.1 for this purpose, too.

|     |     | 1   |     |     |
| --- | --- | --- | --- | --- |
|     | 0   |     | 0   |     |
|     |     | $H^1(\mathcal{M}; T_\mathcal{M}^*)$ |     |     |
|     | $H^2(\mathcal{M}; T_\mathcal{M}^*)$ |     | $H^2(\mathcal{M}; T_\mathcal{M}^*)$ |     |
| 1   |     | $H^1(\mathcal{M}; T_\mathcal{M}^*)$ |     | 1 . |
|     | 0   |     | 0   |     |
|     |     | 0   |     |     |
|     |     | 1   |     |     |

**Table 7.3.:** General Hodge diamond of a Calabi-Yau three-fold

## 7.3 Tangent bundle-valued cohomology

Let us begin basic with a very natural vector bundle of the toric ambient variety $\mathbb{P}_\Sigma$, i.e. its tangent bundle $T_{\mathbb{P}_\Sigma}$. Let $\mathcal{D}_i$ be the toric divisor $\{x_i = 0\}$ associated to each homogeneous coordinate $x_i$ of $\mathbb{P}_\Sigma$. Then there is a short exact sequence

$$0 \longrightarrow \mathcal{O}_{\mathbb{P}_\Sigma}^{\oplus r} \xrightarrow{\otimes Q_i x_i} \bigoplus_{i=1}^{n} \mathcal{O}_{\mathbb{P}_\Sigma}(\mathcal{D}_i) \longrightarrow T_{\mathbb{P}_\Sigma} \longrightarrow 0, \quad \mathcal{O}_{\mathbb{P}_\Sigma}^{\oplus r} := \bigoplus_{k=1}^{r} \mathcal{O}_{\mathbb{P}_\Sigma} \qquad (7.5)$$

where $r$ denotes the number of linear equivalence relations defining the toric variety as before. One can show that (7.5) can be used to define the tangent bundle of $\mathbb{P}_\Sigma$ as the following quotient space

$$T_{\mathbb{P}_\Sigma} = \frac{\bigoplus_{i=1}^{n} \mathcal{O}_{\mathbb{P}_\Sigma}(\mathcal{D}_i)}{\text{im}(Q_i x_i)}. \qquad (7.6)$$

Furthermore, due to exactness, (7.5) provides us with a long exact sequence in cohomology using theorem 4.3.2

$$\begin{aligned}
0 &\longrightarrow H^0(\mathbb{P}_\Sigma; \mathcal{O}_{\mathbb{P}_\Sigma})^{\oplus r} \longrightarrow \bigoplus_{k=1}^{n} H^0(\mathbb{P}_\Sigma; \mathcal{O}_{\mathbb{P}_\Sigma}(\mathcal{D}_k)) \longrightarrow H^0(\mathbb{P}_\Sigma; T_{\mathbb{P}_\Sigma}) \\
&\longrightarrow H^1(\mathbb{P}_\Sigma; \mathcal{O}_{\mathbb{P}_\Sigma})^{\oplus r} \longrightarrow \bigoplus_{k=1}^{n} H^1(\mathbb{P}_\Sigma; \mathcal{O}_{\mathbb{P}_\Sigma}(\mathcal{D}_k)) \longrightarrow H^1(\mathbb{P}_\Sigma; T_{\mathbb{P}_\Sigma}) \\
&\longrightarrow H^2(\mathbb{P}_\Sigma; \mathcal{O}_{\mathbb{P}_\Sigma})^{\oplus r} \longrightarrow \bigoplus_{k=1}^{n} H^2(\mathbb{P}_\Sigma; \mathcal{O}_{\mathbb{P}_\Sigma}(\mathcal{D}_k)) \longrightarrow H^2(\mathbb{P}_\Sigma; T_{\mathbb{P}_\Sigma}) \longrightarrow \ldots
\end{aligned} \qquad (7.7)$$

which allows by computing $H^i(\mathbb{P}_\Sigma; \mathcal{O}_{\mathbb{P}_\Sigma})$ as well as $H^i(\mathbb{P}_\Sigma; \mathcal{O}_{\mathbb{P}_\Sigma}(\mathcal{D}_k))$ the derivation of $H^i(\mathbb{P}_\Sigma; T_{\mathbb{P}_\Sigma})$. The map $\otimes Q_i x_i$ in (7.5) generalizes canonically to a map in (7.7). For the cotangent bundle the story is just as easy. The only thing we have to do

## 7.3. Tangent bundle-valued cohomology

is dualizing (7.5) which gives

$$0 \longrightarrow T^*_{\mathbb{P}_\Sigma} \cong \Omega^1_{\mathbb{P}_\Sigma} \hookrightarrow \bigoplus_{k=1}^{n} \mathcal{O}_{\mathbb{P}_\Sigma}(-\mathcal{D}_k) \twoheadrightarrow \mathcal{O}^{\oplus r}_{\mathbb{P}_\Sigma} \longrightarrow 0, \qquad (7.8)$$

where exactness is valid also in the dual sequence and one may derive the induced long exact sequence in order to calculate $H^i(\mathbb{P}_\Sigma; T^*_{\mathbb{P}_\Sigma})$.

### 7.3.1 Hypersurfaces

So far we only know how to derive the (co)tangent bundle-valued cohomology groups of the toric ambient space. The situation becomes a little bit more involved if we are interested in the (co)tangent bundle of a hypersurface inside some toric variety. Let $\mathcal{S}$ denote the divisor class of a hypersurface $\mathcal{S} = \{x \in \mathbb{P}_\Sigma \mid G = 0\}$. Then the tangent bundle of the subvariety $\mathcal{S}$ can be derived as the measure of non-exactness of the sequence

$$0 \longrightarrow \mathcal{O}^{\oplus r}_\mathcal{S} \xrightarrow{\otimes Q_i x_i} \bigoplus_{i=1}^{n} \mathcal{O}_\mathcal{S}(\mathcal{D}_i) \xrightarrow{\otimes \frac{\partial G}{\partial x_i}} \mathcal{O}_\mathcal{S}(\mathcal{S}) \longrightarrow 0, \qquad (7.9)$$

i.e. with respect to the mappings we have

$$T_\mathcal{S} = \frac{\ker(\frac{\partial G}{\partial x_i})}{\operatorname{im}(Q_i x_i)} \qquad (7.10)$$

which defines a quotient bundle of $\bigoplus_{k=1}^{n} \mathcal{O}_\mathcal{S}(\mathcal{D}_k)$. What we ultimately want is the cohomology of the tangent bundle an hence we need to split this non-exact sequence into short exact sequences. Such a split is always possible once one starts with a short complex and can be derived as pictorially shown in figure 7.1. In this case here we denote the quotient that defines the auxiliary sheaf by $\mathcal{E}_\mathcal{S}$ and hence obtain

$$0 \longrightarrow \mathcal{O}^{\oplus r}_\mathcal{S} \xrightarrow{\otimes x_i} \bigoplus_{k=1}^{n} \mathcal{O}_\mathcal{S}(\mathcal{D}_k) \xrightarrow{R} \mathcal{E}_\mathcal{S} \longrightarrow 0,$$
$$0 \longrightarrow T_\mathcal{S} \xhookrightarrow{\iota} \mathcal{E}_\mathcal{S} \xrightarrow{\otimes \frac{\partial G}{\partial x_i}} \mathcal{O}_\mathcal{S}(\mathcal{S}) \longrightarrow 0, \qquad (7.11)$$

where $R$ is the restriction map to $Q_i x_i = 0$. These are now exact, i.e. we effectively represent the definition of the hypersurface's tangent bundle by a split into two

# 7. Vector Bundle-Valued Cohomology

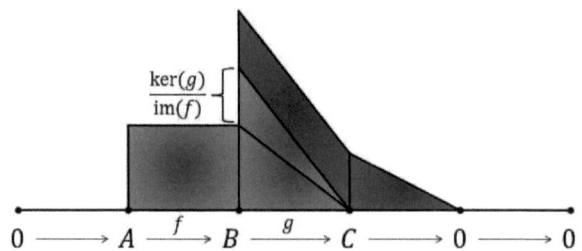

(a) Short complex with cohomology in the middle given as the quotient of the kernel of $g$ by the image of $f$.

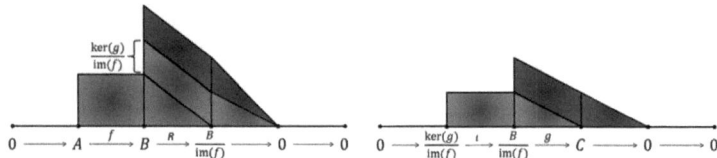

(b) Split of the short complex into two short exact sequences. The cohomology of the complex in figure 7.1a is now the first space in the second exact sequence.

**Figure 7.1.:** Split of a complex with cohomology into two exact sequences.

exact sequences. The auxiliary sheaf $\mathcal{E}_\mathcal{S}$ in (7.11) is just given as the quotient

$$\mathcal{E}_\mathcal{S} := \frac{\bigoplus_{i=1}^{n} \mathcal{O}_\mathcal{S}(\mathcal{D}_i)}{\operatorname{im}(Q_i x_i)}. \tag{7.12}$$

From another perspective the sheaf $\mathcal{E}_\mathcal{S}$ can be identified with the restriction of the tangent bundle sheaf of the ambient space $\mathbb{P}_\Sigma$, i.e. one may treat $\mathcal{E}_\mathcal{S}$ like $T_{\mathbb{P}_\Sigma}|_\mathcal{S}$.

Thus, following the by now established method of using the exactness of the induced long exact cohomology sequences we first may determine the sheaf cohomology $H^i(\mathcal{S}; \mathcal{E}_\mathcal{S})$ from the first sequence and then run through the second sequence to determine $H^i(\mathcal{S}; T_\mathcal{S})$. It is also necessary to compute the restrictions of $\mathcal{O}_{\mathbb{P}_\Sigma}(\mathcal{D}_i)$ to $\mathcal{S}$ which is accomplished via tensoring the Koszul sequence (6.9) with the line bundle $\mathcal{O}_{\mathbb{P}_\Sigma}(\mathcal{D}_i)$, i.e.

$$0 \longrightarrow \underbrace{\mathcal{O}_{\mathbb{P}_\Sigma}(\mathcal{D}_i) \otimes \mathcal{O}_{\mathbb{P}_\Sigma}(-\mathcal{S})}_{\mathcal{O}_{\mathbb{P}_\Sigma}(\mathcal{D}_i - \mathcal{S})} \hookrightarrow \underbrace{\mathcal{O}_{\mathbb{P}_\Sigma}(\mathcal{D}_i) \otimes \mathcal{O}_{\mathbb{P}_\Sigma}}_{\mathcal{O}_{\mathbb{P}_\Sigma}(\mathcal{D}_i)} \longrightarrow \underbrace{\mathcal{O}_{\mathbb{P}_\Sigma}(\mathcal{D}_i) \otimes \mathcal{O}_\mathcal{S}}_{\mathcal{O}_\mathcal{S}(\mathcal{D}_i)} \longrightarrow 0. \tag{7.13}$$

which allows for the computation of the corresponding cohomology groups. In the same way we have to determine the cohomology groups of $\mathcal{O}_\mathcal{S}(\mathcal{S})$ which gives us everything to obtain the desired cohomology groups $H^i(\mathcal{S}; T_\mathcal{S})$.

We have seen in (7.2.1) that the Hodge numbers of a Calabi-Yau three-fold $\mathcal{M}$

## 7.3. Tangent bundle-valued cohomology

can be obtained from the cotangent bundle-valued cohomology groups as

$$h^{1,1}_{\mathcal{M}} = h^1(\mathcal{M}; T^*_{\mathcal{M}}) \qquad h^{1,2}_{\mathcal{M}} = h^2(\mathcal{M}; T^*_{\mathcal{M}}). \tag{7.14}$$

To get the cotangent bundle-valued cohomology groups of any subvariety $\mathcal{S}$ we just have to dualize all the short exact sequences we just used. This yields the four sequences

$$\begin{aligned}
0 &\longrightarrow \mathcal{E}^*_{\mathcal{S}} \hookrightarrow \bigoplus_{i=1}^n \mathcal{O}_{\mathcal{S}}(-\mathcal{D}_i) \longrightarrow \mathcal{O}^{\oplus r}_{\mathcal{S}} \longrightarrow 0, \\
0 &\longrightarrow \mathcal{O}_{\mathcal{S}}(-\mathcal{S}) \hookrightarrow \mathcal{E}^*_{\mathcal{S}} \longrightarrow T^*_{\mathcal{S}} \cong \Omega^1_{\mathcal{S}} \longrightarrow 0, \\
0 &\longrightarrow \mathcal{O}_{\mathbb{P}_\Sigma}(\mathcal{D}_i - \mathcal{S}) \hookrightarrow \mathcal{O}_{\mathbb{P}_\Sigma}(\mathcal{D}_i) \longrightarrow \mathcal{O}_{\mathcal{S}}(\mathcal{D}_i) \longrightarrow 0, \\
0 &\longrightarrow \mathcal{O}_{\mathbb{P}_\Sigma} \hookrightarrow \mathcal{O}_{\mathbb{P}_\Sigma}(\mathcal{D}_i) \longrightarrow \mathcal{O}_{\mathcal{S}}(\mathcal{D}_i) \longrightarrow 0
\end{aligned} \tag{7.15}$$

which we have to run through to get the cotangent bundle-valued cohomology groups.

*Remark.* In fact the cohomology of the tangent bundle and the cotangent bundle are related via

$$H^i(\mathcal{S}; T_{\mathcal{S}}) = H^{d_s - i}(\mathcal{S}; T^*_{\mathcal{S}}). \tag{7.16}$$

Hence it suffices to calculate either one or the other and the other will be given immediately .

**Example 7.3.1** (Hodge diamond of the octic hypersurface in $\mathbb{P}^4_{11222}$). To exemplify the discussed methods, we compute the Hodge diamond of the embedded Calabi-Yau hypersurface $\mathcal{M} := \mathbb{P}^4_{11222}[8]$. Since the weighted projective ambient space $\mathbb{P}^4_{11222}$ has a $\mathbb{Z}_2$-singularity and our methods only apply for smooth spaces, we first perform a toric blow up to get the toric data of the smooth ambient space $\mathbb{P}_\Sigma$ as shown in Table 7.4. We will look at the Calabi-Yau hypersurface $\mathcal{M}$ in $\mathbb{P}_\Sigma$ given by a polynomial of homogeneous degree $\mathcal{M} = (8, 4)$ and compute its Hodge diamond. Inserting the data into the first two sequences of (7.15), we get

$$\begin{aligned}
0 &\longrightarrow \mathcal{E}^*_{\mathcal{M}} \hookrightarrow \mathcal{O}_{\mathcal{M}}(-1, 0)^{\oplus 2} \oplus \mathcal{O}_{\mathcal{M}}(-2, -1)^{\oplus 3} \oplus \mathcal{O}_{\mathcal{M}}(0, -1) \longrightarrow \mathcal{O}^{\oplus 2}_{\mathcal{M}} \longrightarrow 0 \\
0 &\longrightarrow \mathcal{O}_{\mathcal{M}}(-8, -4) \hookrightarrow \mathcal{E}^*_{\mathcal{M}} \longrightarrow T^*_{\mathcal{M}} \cong \Omega^1_{\mathcal{M}} \longrightarrow 0.
\end{aligned} \tag{7.17}$$

To make use of these, it is necessary to determine the cohomology of $\mathcal{O}_{\mathcal{M}}(-1, 0)$, $\mathcal{O}_{\mathcal{M}}(-2, -1)$, $\mathcal{O}_{\mathcal{M}}(0, -1)$, $\mathcal{O}_{\mathcal{M}}$ and $\mathcal{O}_{\mathcal{M}}(-8, -4)$, which is done using the Koszul sequence (6.12). For example, to get $\mathcal{O}_{\mathcal{M}}(-1, 0)$, one takes the short exact se-

| vertices of the polyhedron / fan | coords | GLSM charges $Q^1$ $Q^2$ | | divisor class |
|---|---|---|---|---|
| $v_1 = (-1, -2, -2, -2)$ | $x_1$ | 1 | 0 | $H$ |
| $v_2 = (\phantom{-}1, \phantom{-}0, \phantom{-}0, \phantom{-}0)$ | $x_2$ | 1 | 0 | $H$ |
| $v_3 = (\phantom{-}0, \phantom{-}1, \phantom{-}0, \phantom{-}0)$ | $x_3$ | 2 | 1 | $2H + X$ |
| $v_4 = (\phantom{-}0, \phantom{-}0, \phantom{-}1, \phantom{-}0)$ | $x_4$ | 2 | 1 | $2H + X$ |
| $v_5 = (\phantom{-}0, \phantom{-}0, \phantom{-}0, \phantom{-}1)$ | $x_5$ | 2 | 1 | $2H + X$ |
| $v_6 = (\phantom{-}0, -1, -1, -1)$ | $x_6$ | 0 | 1 | $X$ |
| conditions: | | 8 | 4 | |

$$\mathcal{SR}(\mathbb{P}_\Sigma) = \langle x_1 x_2,\ x_3 x_4 x_5 x_6 \rangle$$
$$\Sigma(\mathbb{P}_\Sigma) = \langle [2\,3\,4\,5],\ [1\,3\,4\,5],\ [2\,3\,4\,6],\ [2\,4\,5\,6],$$
$$[2\,3\,5\,6],\ [1\,3\,4\,6],\ [1\,4\,5\,6],\ [1\,3\,5\,6] \rangle$$

**Table 7.4.:** The torically blown-up weighted projective space $\mathbb{P}^4_{11222}$ to its smooth version $\mathbb{P}_\Sigma$ with embedded Calabi-Yau hypersurface given by the divisor $8H + 4X$ with charges $(8, 4)$.

quence

$$0 \longrightarrow \mathcal{O}_{\mathbb{P}_\Sigma}(-9, -4) \hookrightarrow \mathcal{O}_{\mathbb{P}_\Sigma}(-1, 0) \longrightarrow \mathcal{O}_\mathcal{M}(-1, 0) \longrightarrow 0 \tag{7.18}$$

and then looks at the long exact sequence in cohomology. Therefore, it is sufficient to know the cohomology of the ambient space line bundles $\mathcal{O}_{\mathbb{P}_\Sigma}(\mathcal{D})$ with divisor charges

$$\mathcal{D} \in \big\{ (-9, -4), (-1, 0), (-10, -5), (-2, -1), \\ (-8, -5), (0, -1), (-8, -4), (0, 0), (-16, 8) \big\}, \tag{7.19}$$

for which our algorithm yields the cohomology group dimensions

$$h^\bullet(\mathbb{P}_\Sigma; \mathcal{O}_{\mathbb{P}_\Sigma}(-9, -4)) = (0, 0, 0, 0, 2), \qquad h^\bullet(\mathbb{P}_\Sigma; \mathcal{O}_{\mathbb{P}_\Sigma}(-1, 0)) = (0, 0, 0, 0, 0),$$
$$h^\bullet(\mathbb{P}_\Sigma; \mathcal{O}_{\mathbb{P}_\Sigma}(-10, -5)) = (0, 0, 0, 0, 6), \qquad h^\bullet(\mathbb{P}_\Sigma; \mathcal{O}_{\mathbb{P}_\Sigma}(-2, -1)) = (0, 0, 0, 0, 0),$$
$$h^\bullet(\mathbb{P}_\Sigma; \mathcal{O}_{\mathbb{P}_\Sigma}(-8, -5)) = (0, 0, 0, 3, 1), \qquad h^\bullet(\mathbb{P}_\Sigma; \mathcal{O}_{\mathbb{P}_\Sigma}(0, -1)) = (0, 0, 0, 0, 0),$$
$$h^\bullet(\mathbb{P}_\Sigma; \mathcal{O}_{\mathbb{P}_\Sigma}(-8, -4)) = (0, 0, 0, 0, 1), \qquad h^\bullet(\mathbb{P}_\Sigma; \mathcal{O}_{\mathbb{P}_\Sigma}(0, 0)) = (1, 0, 0, 0, 0),$$
$$h^\bullet(\mathbb{P}_\Sigma; \mathcal{O}_{\mathbb{P}_\Sigma}(-16, -8)) = (0, 0, 0, 0, 105). \tag{7.20}$$

Note already this extra contribution $h^3(\mathbb{P}_\Sigma; \mathcal{O}_{\mathbb{P}_\Sigma}(-8, -5)) = 3$ which we will discuss in a moment. Without the explicit knowledge about the generic hypersurface equation $G = 0$ we can obtain the dimensions of the cohomology groups of these

## 7.3. Tangent bundle-valued cohomology

line bundles over $\mathcal{M}$

$$h^\bullet(\mathcal{M}; \mathcal{O}_\mathcal{M}(-1,0)) = (0,0,0,2), \qquad h^\bullet(\mathcal{M}; \mathcal{O}_\mathcal{M}(-2,-1)) = (0,0,0,6),$$
$$h^\bullet(\mathcal{M}; \mathcal{O}_\mathcal{M}(0,-1)) = (0,0,3,1), \qquad h^\bullet(\mathcal{M}; \mathcal{O}_\mathcal{M}(0,0)) = (1,0,0,1),$$
$$h^\bullet(\mathcal{M}; \mathcal{O}_\mathcal{M}(-8,-4)) = (0,0,0,104). \tag{7.21}$$

Likewise, we use those dimensions to determine the cohomology of the auxiliary bundle $\mathcal{E}_\mathcal{M}^*$ from the first sequence of (7.17), yielding

$$h^\bullet(\mathcal{M}; \mathcal{E}_\mathcal{M}^*) = (0,2,3,21). \tag{7.22}$$

In order to compute the cohomology of $\Omega_\mathcal{M}^1$ from the second and final sequence in (7.17) some small additional input is required. Using (7.21) and (7.22), the induced long exact sequence takes the form

$$\begin{array}{c}
0 \longrightarrow \mathbb{R}^0 = 0 \longrightarrow 0 \longrightarrow H^0(\mathcal{M}; \Omega_\mathcal{M}^1) \longrightarrow \\
\longrightarrow 0 \longrightarrow \mathbb{R}^2 \longrightarrow H^1(\mathcal{M}; \Omega_\mathcal{M}^1) \longrightarrow \\
\longrightarrow 0 \longrightarrow \mathbb{R}^3 \longrightarrow H^2(\mathcal{M}; \Omega_\mathcal{M}^1) \longrightarrow \\
\longrightarrow \mathbb{R}^{104} \longrightarrow \mathbb{R}^{21} \longrightarrow H^3(\mathcal{M}; \Omega_\mathcal{M}^1) \longrightarrow 0.
\end{array} \tag{7.23}$$

Whereas $H^0(\mathcal{M}; \Omega_\mathcal{M}^1) = 0$ and $H^1(\mathcal{M}; \Omega_\mathcal{M}^1) = \mathbb{R}^2$ follow immediately from the sequence, the remaining two cohomology groups seem to be ambiguous. However, one should keep in mind that via the symmetries in the Hodge diamond it follows that

$$H^3(\mathcal{M}; \Omega_\mathcal{M}^1) \cong H^{0,2}(\mathcal{M}) \cong H^2(\mathcal{M}; \mathcal{O}_\mathcal{M}) = 0. \tag{7.24}$$

The remaining part of the sequence therefore reads

$$0 \longrightarrow \mathbb{R}^3 \hookrightarrow H^2(\mathcal{M}; \Omega_\mathcal{M}^1) \longrightarrow \mathbb{R}^{104} \twoheadrightarrow \mathbb{R}^{21} \longrightarrow 0, \tag{7.25}$$

such that via $3 - \dim H^2(\mathcal{M}; \Omega_\mathcal{M}^1) + 104 - 21 = 0$ as required for exactness we can determine the result

$$h^\bullet(\Omega_\mathcal{M}^1) = (0,2,86,0). \tag{7.26}$$

This ultimately gives us the Hodge diamond

$$
\begin{array}{ccccccc}
 & & & h^{0,0} & & & \\
 & & h^{1,0} & & h^{0,1} & & \\
 & h^{2,0} & & h^{1,1} & & h^{0,2} & \\
h^{3,0} & & h^{2,1} & & h^{1,2} & & h^{0,3} \\
 & h^{2,0} & & h^{1,1} & & h^{0,2} & \\
 & & h^{1,0} & & h^{0,1} & & \\
 & & & h^{0,0} & & &
\end{array}
=
\begin{array}{ccccccc}
 & & & 1 & & & \\
 & & 0 & & 0 & & \\
 & 0 & & 2 & & 0 & \\
1 & & 86 & & 86 & & 1 \\
 & 0 & & 2 & & 0 & \\
 & & 0 & & 0 & & \\
 & & & 1 & & &
\end{array}
\quad
\begin{array}{l}
b^0 = 1 \\
b^1 = 0 \\
b^2 = 2 \\
b^3 = 174 \\
b^4 = 2 \\
b^5 = 0 \\
b^6 = 1
\end{array}
\quad (7.27)
$$

for the octic Calabi-Yau three-fold hypersurface $\mathbb{P}^4_{11222}[8]$. In prospect of chapter 8 we introduce[1] the numbers $\mathfrak{h}^{p,q}_i$ which refer to the contribution from the $i^{\text{th}}$ line bundle cohomology group $H^i(\mathbb{P}_\Sigma; \mathcal{O}_{\mathbb{P}_\Sigma}(m,n))$ to the Hodge number $h^{p,q}$:

$$h^i(\mathbb{P}_\Sigma; \mathcal{O}_{\mathbb{P}_\Sigma}(m,n)) \rightsquigarrow \mathfrak{h}^{p,q}_i. \qquad (7.28)$$

It is clear from (7.23) that $H^{2,1}(\mathcal{M}) = H^2(\mathcal{M}; \Omega^1_\mathcal{M})$ receives two different kinds of contributions, $\mathfrak{h}^{2,1}_4(\mathcal{M}) = 83$ coming from elements in $H^4(\mathbb{P}_\Sigma, \mathcal{O}_{\mathbb{P}_\Sigma}(m,n))$ and $\mathfrak{h}^{2,1}_3(\mathcal{M}) = 3$ from elements in $H^3(\mathbb{P}_\Sigma, \mathcal{O}_{\mathbb{P}_\Sigma}(m,n))$[2] As we will see in chapter 8, these contributions are also related to the non-geometric contributions in the Batyrev formula.

### 7.3.2 Complete intersection subvarieties

In the last subsection we described a method to calculate the dimensions of the tangent bundle-valued cohomology groups as well as the Hodge numbers for three-dimensional hypersurfaces. Now we want to generalize these methods to the case where the subvariety does not arise as a hypersurface of a toric variety, but rather as a complete intersection of several hypersurfaces. Conceptually it is not much of a difference and we proceed in the same way as for the case of a single hypersurface.

Let $\{\mathcal{S}_1, \ldots, \mathcal{S}_c\}$ be a set of hypersurfaces in a toric variety $\mathbb{P}_\Sigma$ such that their complete intersection subvariety is of codimension $c$ and denoted by $\mathcal{S}$. The tangent bundle of $\mathcal{S}$ is then, in analogy to (7.9), given by the cohomology of the

---

[1] In order to avoid any confusion, note that the $\mathfrak{h}^{p,q}_i$ here are entirely unrelated to the multiplicity factors $\mathfrak{h}^i(\mathcal{Q})$ of chapter 4.

[2] Let us mention that this split can also be seen in the corresponding Gepner and Landau-Ginzburg orbifold models, where precisely 3 massless matter states come from so-called twisted sectors and the remaining 83 from the untwisted sector.

## 7.3. Tangent bundle-valued cohomology

complex

$$0 \longrightarrow \mathcal{O}_{\mathcal{S}}^{\oplus r} \xrightarrow{\otimes Q_i x_i} \bigoplus_{i=1}^{s} \mathcal{O}_{\mathcal{S}}(\mathcal{D}_i) \xrightarrow{\frac{\partial G_j}{\partial x_i}} \bigoplus_{j=1}^{c} \mathcal{O}_{\mathcal{S}}(\mathcal{S}_j) \longrightarrow 0, \qquad (7.29)$$

where as before the $\mathcal{D}_i$ denote the toric divisors for the respective coordinates. As before we can perform a splitting of this complex into the two exact sequences

$$\begin{aligned} 0 &\longrightarrow \mathcal{O}_{\mathcal{S}}^{\oplus r} \hookrightarrow \bigoplus_{i=1}^{n} \mathcal{O}_{\mathcal{S}}(\mathcal{D}_i) \longrightarrow \mathcal{E}_{\mathcal{S}} \longrightarrow 0, ] \\ 0 &\longrightarrow T_{\mathcal{S}} \hookrightarrow \mathcal{E}_{\mathcal{S}} \longrightarrow \bigoplus_{j=1}^{l} \mathcal{O}_{\mathcal{S}}(\mathcal{S}_j) \longrightarrow 0. \end{aligned} \qquad (7.30)$$

The dual version for the cotangent bundle reads

$$\begin{aligned} 0 &\longrightarrow \mathcal{E}_{\mathcal{S}}^* \hookrightarrow \bigoplus_{k=1}^{n} \mathcal{O}_{\mathcal{S}}(-\mathcal{D}_k) \longrightarrow \mathcal{O}_{\mathcal{S}}^{\oplus r} \longrightarrow 0, \\ 0 &\longrightarrow \bigoplus_{i=1}^{l} \mathcal{O}_{\mathcal{S}}(-\mathcal{S}_i) \hookrightarrow \mathcal{E}_{\mathcal{S}}^* \longrightarrow \Omega_{\mathcal{S}}^1 \longrightarrow 0 \end{aligned} \qquad (7.31)$$

which of course for $c = 1$ precisely reproduces the hypersurface result (7.11).

*Remark.* The generalization of codimension one to arbitrary codimensional subvarieties seems to be fairly easy since the only thing we have to do is to calculate one more line bundle in (7.31) or (7.29) for each hypersurface. Conceptually this really is the only difference but technically this is a huge difference. While for the hypersurface we could employ the short Koszul sequence (6.12) for every line bundle we want to obtain, for the case of an intersection of more than one hypersurface we have to use the long Koszul sequence (6.31) which produces one more short exact sequence for each additional hypersurface and is also technically harder to resolve. The higher the codimension the more involved the calculation and in fact a computer implementation of these calculations are quite early unavoidable.

**Example 7.3.2** (Hodge diamond of the double quartic in $\mathbb{P}^5_{111122}$). Let us now see how the methods for a complete intersection Calabi-Yau work in detail for the case of two intersecting hypersurfaces, by examining the specific case of $\mathcal{M} := \mathbb{P}^5_{111122}[4,4]$ living in the weighted projective ambient space $\mathbb{P}_\Sigma := \mathbb{P}^5_{111122}$. The corresponding toric data is given in table 7.5. For this example the sequences

| vertices of the polyhedron / fan | coords | GLSM charges $Q^1$ | divisor class |
|---|---|---|---|
| $v_1 = (\,-1,\ -1,\ -1,\ -2,\ -2\,)$ | $x_1$ | 1 | $H$ |
| $v_2 = (\ \ \,1,\ \ \ 0,\ \ \ 0,\ \ \ 0,\ \ \ 0\,)$ | $x_2$ | 1 | $H$ |
| $v_3 = (\ \ \,0,\ \ \ 1,\ \ \ 0,\ \ \ 0,\ \ \ 0\,)$ | $x_3$ | 1 | $H$ |
| $v_4 = (\ \ \,0,\ \ \ 0,\ \ \ 1,\ \ \ 0,\ \ \ 0\,)$ | $x_4$ | 1 | $H$ |
| $v_5 = (\ \ \,0,\ \ \ 0,\ \ \ 0,\ \ \ 1,\ \ \ 0\,)$ | $x_5$ | 2 | $2H$ |
| $v_6 = (\ \ \,0,\ \ \ 0,\ \ \ 0,\ \ \ 0,\ \ \ 1\,)$ | $x_6$ | 2 | $2H$ |
| conditions: | | 4 | |
| | | 4 | |

$$\text{intersection form: } \tfrac{1}{4}H^5$$
$$\mathrm{SR}(\mathbb{P}_\Sigma) = \langle x_1 x_2 x_3 x_4 x_5 x_6 \rangle$$

**Table 7.5.:** Toric data for the complete intersection Calabi-Yau threefold $\mathcal{M} := \mathbb{P}^5_{111122}[4,4]$ living in the ambient space $\mathbb{P}_\Sigma := \mathbb{P}^5_{111122}$.

(7.31) reduce to

$$
\begin{aligned}
0 &\longrightarrow \mathcal{E}^*_\mathcal{M} \hookrightarrow \mathcal{O}_\mathcal{M}(-1)^{\oplus 4} \oplus \mathcal{O}_\mathcal{M}(-2)^{\oplus 2} \twoheadrightarrow \mathcal{O}_\mathcal{M} \longrightarrow 0\,, \\
0 &\longrightarrow \mathcal{O}_\mathcal{M}(-4) \oplus \mathcal{O}_\mathcal{M}(-4) \hookrightarrow \mathcal{E}^*_\mathcal{M} \twoheadrightarrow \Omega_\mathcal{M} \longrightarrow 0
\end{aligned}
\qquad (7.32)
$$

and hence we need to determine the cohomologies of the line bundles $\mathcal{O}_\mathcal{M}(-1)$, $\mathcal{O}_\mathcal{M}(-2)$, $\mathcal{O}_\mathcal{M}$, $\mathcal{O}_\mathcal{M}(-4)$ over the complete intersection Calabi-Yau. This can be done by employing equations (6.38) which give us the four pairs of sequences, one pair for each line bundle

$$
\begin{aligned}
0 &\longrightarrow \mathcal{O}_{\mathbb{P}_\Sigma}(-9) \hookrightarrow \mathcal{O}_{\mathbb{P}_\Sigma}(-5)^{\oplus 2} \twoheadrightarrow \mathcal{I}_a \longrightarrow 0\,, \\
0 &\longrightarrow \mathcal{I}_a \hookrightarrow \mathcal{O}_{\mathbb{P}_\Sigma}(-1) \twoheadrightarrow \mathcal{O}_\mathcal{M}(-1) \longrightarrow 0\,,
\end{aligned}
\qquad (7.33)
$$

$$
\begin{aligned}
0 &\longrightarrow \mathcal{O}_{\mathbb{P}_\Sigma}(-10) \hookrightarrow \mathcal{O}_{\mathbb{P}_\Sigma}(-6)^{\oplus 2} \twoheadrightarrow \mathcal{I}_b \longrightarrow 0\,, \\
0 &\longrightarrow \mathcal{I}_b \hookrightarrow \mathcal{O}_{\mathbb{P}_\Sigma}(-2) \twoheadrightarrow \mathcal{O}_\mathcal{M}(-2) \longrightarrow 0\,,
\end{aligned}
\qquad (7.34)
$$

$$
\begin{aligned}
0 &\longrightarrow \mathcal{O}_{\mathbb{P}_\Sigma}(-8) \hookrightarrow \mathcal{O}_{\mathbb{P}_\Sigma}(-4)^{\oplus 2} \twoheadrightarrow \mathcal{I}_c \longrightarrow 0\,, \\
0 &\longrightarrow \mathcal{I}_c \hookrightarrow \mathcal{O}_{\mathbb{P}_\Sigma} \twoheadrightarrow \mathcal{O}_\mathcal{M} \longrightarrow 0\,,
\end{aligned}
\qquad (7.35)
$$

## 7.3. Tangent bundle-valued cohomology

$$0 \longrightarrow \mathcal{O}_{\mathbb{P}_\Sigma}(-12) \hookrightarrow \mathcal{O}_{\mathbb{P}_\Sigma}(-8)^{\oplus 2} \longrightarrow \mathcal{I}_d \longrightarrow 0\,,$$
$$0 \longrightarrow \mathcal{I}_d \hookrightarrow \mathcal{O}_{\mathbb{P}_\Sigma}(-4) \longrightarrow \mathcal{O}_\mathcal{M}(-4) \longrightarrow 0\,. \qquad (7.36)$$

Deriving the corresponding long exact sequences of the cohomology groups allows us to determine for each pair first the dimensions of the cohomologies of the auxiliary sheaf and then in the second step the one for the line bundle itself. For some of the line bundles in (7.33)-(7.36) all cohomology groups vanish. For those where this is not the case we find

$$h^\bullet(\mathbb{P}_\Sigma; \mathcal{O}_{\mathbb{P}_\Sigma}(-9)) = (0,0,0,0,0,4)\,, \quad h^\bullet(\mathbb{P}_\Sigma; \mathcal{O}_{\mathbb{P}_\Sigma}(-10)) = (0,0,0,0,0,12)\,,$$
$$h^\bullet(\mathbb{P}_\Sigma; \mathcal{O}_{\mathbb{P}_\Sigma}(-8)) = (0,0,0,0,0,1)\,, \quad h^\bullet(\mathbb{P}_\Sigma; \mathcal{O}_{\mathbb{P}_\Sigma}(-12)) = (0,0,0,0,0,58)\,,$$
$$h^\bullet(\mathbb{P}_\Sigma; \mathcal{O}_{\mathbb{P}_\Sigma}) = (1,0,0,0,0,0)\,, \qquad (7.37)$$

from which follows the cohomology of the auxiliary sheaves where this is not the case we find

$$h^\bullet(\mathbb{P}_\Sigma; \mathcal{I}_a) = (0,0,0,0,4,0)\,, \quad h^\bullet(\mathbb{P}_\Sigma; \mathcal{I}_b) = (0,0,0,0,12,0)\,,$$
$$h^\bullet(\mathbb{P}_\Sigma; \mathcal{I}_c) = (0,0,0,0,1,0)\,, \quad h^\bullet(\mathbb{P}_\Sigma; \mathcal{I}_d) = (0,0,0,0,56,0)\,. \qquad (7.38)$$

Taking this into account one can use the second sequences from (7.33)-(7.34) to read off

$$h^\bullet(\mathcal{M}; \mathcal{O}_\mathcal{M}(-1)) = (0,0,0,4)\,, \qquad h^\bullet(\mathcal{M}; \mathcal{O}_\mathcal{M}(-2)) = (0,0,0,12)\,,$$
$$h^\bullet(\mathcal{M}; \mathcal{O}_\mathcal{M}) = (1,0,0,1)\,, \qquad h^\bullet(\mathcal{M}; \mathcal{O}_\mathcal{M}(-4)) = (0,0,0,56)\,, \qquad (7.39)$$

where $h^\bullet(\mathcal{M}; \mathcal{O}_\mathcal{M})$ already presents the expected first row of the Hodge diamond. Now we can proceed in the same way as we did in the last subsection and plug this into the first equation of (7.32) to get

$$h^\bullet(\mathcal{M}; \mathcal{E}_\mathcal{M}^*) = (0,1,0,39)\,. \qquad (7.40)$$

We insert this result together with $h^\bullet(\mathcal{M}; \mathcal{O}_\mathcal{M}(-4))$ from equations (7.39) into the second equation in (7.32). In order to derive a unique result from the long exact sequence, we have to use the fact that the complete intersection is Calabi-Yau which implies that $h^0(\mathcal{M}; \Omega^1_\mathcal{M}) = 0$ and find

$$h^\bullet(\mathcal{M}; \Omega^1_\mathcal{M}) = (0,1,73,0) \qquad (7.41)$$

Since this is the second row of the Hodge diamond we are looking for and since

$\mathcal{M}$ is Calabi-Yau, we can write down the full Hodge diamond as

$$\begin{array}{ccccccc} & & & 1 & & & \\ & & 0 & & 0 & & \\ & 0 & & 1 & & 0 & \\ 1 & & 73 & & 73 & & 1 \\ & 0 & & 1 & & 0 & \\ & & 0 & & 0 & & \\ & & & 1 & & & \end{array} \left| \begin{array}{l} b^0 = 1 \\ b^1 = 0 \\ b^2 = 1 \\ b^3 = 148 \\ b^4 = 1 \\ b^5 = 0 \\ b^6 = 1 \end{array} \right. . \qquad (7.42)$$

Note again that by no means we are using any properties special to this geometry, i.e. the described procedure is completely algorithmic and can be analogously applied to any other setting as long as enough zeros appear in the cohomologies to make use of exactness. All the laborious and somewhat confusing steps involved in this computation can be easily carried out with our programm **cohomCalg Koszul** extension [69] which automates precisely the steps outlined above. In the next section we will give an explicit example where the mapping actually cannot be avoided.

## 7.4 Vector bundle-valued cohomology

In this section we will generalize the methods reviewed in the last section to more general vector bundles that can be constructed via the so-called *monad construction* or *extensions*. These methods actually allow for an algorithmic implementation of the determination of most of the topological data which is again based on the algorithm we presented in theorem 4.6.1. We also wrote the implementation for this and made the available in the just mentioned **cohomCalg Koszul** extension [69].

### 7.4.1 The monad construction

Exact monad

We begin with the easiest monad which describes in fact a holomorphic vector bundle $\mathcal{V}$ which is in some way even simpler than the (co)tangent bundle. When we defined the (co)tangent bundle we made use of the Euler sequence (7.9). Here we had to split it into two parts where the first one was basically the tangent bundle of the ambient space restricted to the subvariety (7.30). The idea of an exact monad is to use precisely use this kind of short exact sequence to define a different vector bundle than the tangent bundle. Therefore the line bundles that

## 7.4. Vector bundle-valued cohomology

appear in such a monad will in general have nothing to do with the ones that appear in the Euler sequence. Most generally for such a case we can define the vector bundle by

$$0 \longrightarrow \mathcal{V} \overset{\iota}{\hookrightarrow} \bigoplus_{a=1}^{n_\Lambda} \mathcal{O}_S(N_a) \overset{\otimes F_a{}^l}{\longrightarrow} \bigoplus_{l=1}^{n_p} \mathcal{O}_S(M_l) \longrightarrow 0 \qquad (7.43)$$

where $F_a{}^l$ is a matrix containing polynomials of the right multi-degree, i.e.

$$||F_a{}^l|| = N_a - M_l, \qquad (7.44)$$

and $\iota$ is the inclusion map. Hence this definition of the bundle $\mathcal{V}$ can also be read as

$$\mathcal{V} := \ker\left(F_a{}^l\right). \qquad (7.45)$$

Such a vector bundle will obtain a rank corresponding to the number of line bundles involved in (7.43)

$$\mathrm{rk}(\mathcal{V}) = n_\Lambda - n_p. \qquad (7.46)$$

The choice of the subscripts here are due to the physical interpretation of these numbers and will become clear in chapter 2. Dualizing this sequence gives us the dual version $\mathcal{V}^*$ of $\mathcal{V}$ and it is rather a matter of convention which one is called the dual bundle and which the original. In our convention $\mathcal{V}^*$ is determined by the sequence

$$0 \longrightarrow \bigoplus_{l=1}^{n_p} \mathcal{O}_S(M_l) \overset{\otimes F_a{}^l}{\longrightarrow} \bigoplus_{a=1}^{n_\Lambda} \mathcal{O}_S(N_a) \overset{R}{\longrightarrow} \mathcal{V}^* \longrightarrow 0, \qquad (7.47)$$

with the restriction map $R$ which defines the dual bundle by

$$\mathcal{V}^* := \mathrm{im}\, R = \frac{\bigoplus_{a=1}^{n_\Lambda} \mathcal{O}_S(N_a)}{\mathrm{im}\,(F_a{}^l)}. \qquad (7.48)$$

For the tangent bundle the mappings in the monad were given as multiplication by the coordinate as well as by the derivatives of the hypersurfaces. Here we are basically free to choose the maps $F_a{}^l$ that are part of the definition of the bundle. Neglecting necessities related to the smoothness of the bundle for the moment, the only condition on these maps is (7.44) and we are free to choose them as we please.

**Example 7.4.1** (A positive monad over the quintic). Let us state a fairly easy example especially to see why the maps become crucial for honest calculations of vector bundle-valued cohomology groups. The example is taken from a scan that was performed by the authors of [55, 106]. The Calabi-Yau base manifold is the generic quintic hypersurface in $\mathbb{P}^4$

$$\mathcal{M} = \left\{ x \in \mathbb{P}^4 \mid G = \sum_{\{i_j \mid \sum i_j = 5\}} A_{i_1, i_2, i_3, i_4, i_5} x_1^{i_1} x_2^{i_2} x_3^{i_3} x_4^{i_4} x_5^{i_5} = 0 \right\}. \quad (7.49)$$

and the bundle is defined via the short exact sequence

$$0 \longrightarrow \mathcal{V} \xhookrightarrow{\iota} \mathcal{O}_{\mathcal{M}}^{\oplus 6}(1) \xrightarrow{\otimes F_a^l} \mathcal{O}_{\mathcal{M}}(2)^{\oplus 3} \longrightarrow 0, \quad (7.50)$$

wherein the mapping $F_a^l$ is an $3 \times 6$ matrix that has identical components

$$F_a^l = x_1 + x_2 + x_3 + x_4 + x_5, \quad \forall a i = 1, ... 3 \text{ and } l = 1, ..., 6. \quad (7.51)$$

Since (7.50) is a short exact sequence of sheaves, we can write down a long exact sequence in cohomology 4.3.2. Using the methods from section 6.3.1, we find that

$$H^i(\mathcal{M}; \mathcal{O}_{\mathcal{M}}(1)) = H^i(\mathcal{M}; \mathcal{O}_{\mathcal{M}}(2)) = 0, \quad \forall i \neq 0 \quad (7.52)$$

and the long exact sequence reduces to

$$\begin{array}{l}
0 \longrightarrow H^0(\mathcal{M}; \mathcal{V}) \xrightarrow{\iota} H^0(\mathcal{M}; \mathcal{O}_{\mathcal{M}}(1))^{\oplus 6} \xrightarrow{F_a^l} H^0(\mathcal{M}; \mathcal{O}_{\mathcal{M}}(2))^{\oplus 3} \xrightarrow{\delta} \\
\longrightarrow H^1(\mathcal{M}; \mathcal{V}) \longrightarrow 0 \longrightarrow 0 \longrightarrow \\
\longrightarrow H^2(\mathcal{M}; \mathcal{V}) \longrightarrow 0 \longrightarrow 0 \longrightarrow \\
\longrightarrow H^3(\mathcal{M}; \mathcal{V}) \longrightarrow 0 \longrightarrow 0 \longrightarrow \cdots
\end{array} \quad (7.53)$$

Obviously we find

$$H^2(\mathcal{M}; \mathcal{V}) = H^3(\mathcal{M}; \mathcal{V}) = 0 \quad (7.54)$$

and are left with the task of determining the kernel of $F_a^l$ which will be identical with $H^0(\mathcal{M}; \mathcal{V})$ and then the image of the coboundary map $\delta$ which will give us $H^1(\mathcal{M}; \mathcal{V})$. Since

$$H^0(\mathcal{M}; \mathcal{O}_{\mathcal{M}}(1)) = \langle x_1, x_2, x_3, x_4, x_5 \rangle \text{ and} \quad (7.55)$$

$$H^0(\mathcal{M}; \mathcal{O}_{\mathcal{M}}(2)) = \langle x_i x_j \mid i, j = 1, ..., 5 \rangle \quad (7.56)$$

## 7.4. Vector bundle-valued cohomology

which means that these spaces contain really all sections, there is no way for the map $F_a{}^l$ to send any generator of (7.55) to zero and hence its kernel is simply the zero element

$$\ker(F_a{}^l) = H^0(\mathcal{M}; \mathcal{V}) = 0. \tag{7.57}$$

The remaining space $H^1(\mathcal{M}; \mathcal{V})$ must therefore be isomorphic to the complement of the image of $F_a{}^l$. Since

$$h^0(\mathcal{M}; \mathcal{O}_\mathcal{M}(1)) = 30 \quad \text{and} \quad h^0(\mathcal{M}; \mathcal{O}_\mathcal{M}(2)) = 45, \tag{7.58}$$

we finally find

$$h^1(\mathcal{M}; \mathcal{V}) = 15 \tag{7.59}$$

and hence

$$h^\bullet(\mathcal{M}; \mathcal{V}) = (0, 15, 0, 0). \tag{7.60}$$


Non-exact monad

Similarly to the above exact monad we can now go ahead and simply use a non-exact complex in order to define a holomorphic vector bundle:

$$0 \longrightarrow \bigoplus_{i=1}^{n_F} \mathcal{O}_\mathcal{S}(L_i) \xrightarrow{\otimes E_i{}^a} \bigoplus_{a=1}^{n_\Lambda} \mathcal{O}_\mathcal{S}(N_a) \xrightarrow{\otimes F_a{}^l} \bigoplus_{l=1}^{n_p} \mathcal{O}_\mathcal{S}(M_l) \longrightarrow 0. \tag{7.61}$$

Analogously to the Euler sequence (7.29), the vector bundle $\mathcal{V}$ is then given by the cohomology of this sequence i.e.

$$\mathcal{V} = \frac{\ker(F_a{}^l)}{\operatorname{im}(E_i{}^a)} \tag{7.62}$$

and hence the rank reduces by the number of vector bundles in the first direct sum in (7.61) and is given by

$$\operatorname{rk}(\mathcal{V}) = n_\Lambda - n_p - n_F. \tag{7.63}$$

In contrast to the Euler sequence, we are in principle free to choose the line bundles $L_i$, $N_a$, $M_l$ in (7.61) arbitrarily. Furthermore we are free to choose the mappings in (7.61) as we like as long as they have the right degree, i.e.

$$||E_i{}^a|| = ||N_a|| - ||L_i|| \quad \text{and} \quad ||F_a{}^l|| = ||M_l|| - ||N_a|| \tag{7.64}$$

# 146  7. Vector Bundle-Valued Cohomology

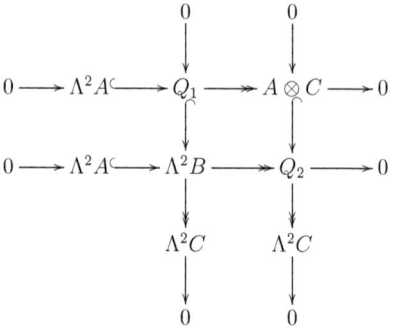

**Table 7.6.:** Sequences to determine $\Lambda^2 A, \Lambda^2 B$ and $\Lambda^2 C$

and make sure that the vector bundle is smooth. The split into two exact sequences is analogous to the Euler sequence:

$$\begin{aligned} 0 \longrightarrow \bigoplus_{i=1}^{n_F} \mathcal{O}_\mathcal{S}(L_i) \hookrightarrow \bigoplus_{a=1}^{n_\Lambda} \mathcal{O}_\mathcal{S}(N_a) \longrightarrow \mathcal{E}_\mathcal{S} \longrightarrow 0, \\ 0 \longrightarrow \mathcal{V} \hookrightarrow \mathcal{E}_\mathcal{S} \twoheadrightarrow \bigoplus_{l=1}^{n_p} \mathcal{O}_\mathcal{S}(M_l) \longrightarrow 0. \end{aligned} \qquad (7.65)$$

## 7.4.2 Cohomology of $\Lambda^2 \mathcal{V}$-, $\Lambda^2 \mathcal{V}^*$

In order to compute the Hodge diamond for higher-dimensional spaces, we need higher exterior powers of the cotangent sheaf, i.e. we require $\Omega^p_{\mathbb{P}_\Sigma}$ for $p > 1$, since the cotangent bundle is in these cases not only of rank three but of higher rank. In chapter 2 we will furthermore see that for heterotic models on Calabi-Yau three-folds equipped with higher rank vector bundle the higher exterior power $\Lambda^k \mathcal{V}$ will become important since they contain information about the spectrum that cannot be obtained by the plain $\mathcal{V}$ bundle-valued cohomology.

### Exact monad

Let us begin with the easier case of an exact monad and to get the idea how it works, consider the following exact sequence

$$0 \longrightarrow A \hookrightarrow B \twoheadrightarrow C \longrightarrow 0 \qquad (7.66)$$

of vector bundles or sheaves. Then all four of the sequences in table 7.6 are short and exact as well: This basically yields two ways to compute $\Lambda^2 A$, $\Lambda^2 B$ and $\Lambda^2 C$

## 7.4. Vector bundle-valued cohomology

using the two split short exact sequences

$$
\begin{aligned}
0 &\longrightarrow \Lambda^2 A \hookrightarrow Q_1 \twoheadrightarrow A \otimes C \longrightarrow 0, \\
0 &\longrightarrow Q_1 \hookrightarrow \Lambda^2 B \twoheadrightarrow \Lambda^2 C \longrightarrow 0
\end{aligned}
\tag{7.67}
$$

or the second pair

$$
\begin{aligned}
0 &\longrightarrow \Lambda^2 A \hookrightarrow \Lambda^2 B \twoheadrightarrow Q_2 \longrightarrow 0, \\
0 &\longrightarrow A \otimes C \hookrightarrow Q_2 \twoheadrightarrow \Lambda^2 C \longrightarrow 0.
\end{aligned}
\tag{7.68}
$$

If we know everything but $A$, the remaining bundle $A \otimes C$ can be obtained by tensoring the exact sequence 7.66 by $C$:

$$
0 \longrightarrow A \otimes C \hookrightarrow B \otimes C \twoheadrightarrow C \otimes C \longrightarrow 0
\tag{7.69}
$$

which has to be taken into account for both ways of the calculation (7.67) and (7.67).

This general approach can now be applied to the exact monad sequence (7.43), such that $A = \mathcal{V}$, $B = \bigoplus_{a=1}^{n_\Lambda} \mathcal{O}_\mathcal{S}(N_a)$ and $C = \bigoplus_{l=1}^{n_p} \mathcal{O}_\mathcal{S}(M_l)$ are used. The necessary sequences (7.67), (7.68) and (7.69) become

$$
\begin{aligned}
0 &\longrightarrow \underaccent{\tilde}{\Lambda^2 \mathcal{V}} \hookrightarrow Q_1 \longrightarrow \mathcal{V} \otimes \bigoplus_{l=1}^{n_p} \mathcal{O}_\mathcal{S}(M_l) \longrightarrow 0, \\
0 &\longrightarrow \underaccent{\tilde}{Q_1} \hookrightarrow \bigoplus_{a_1 < a_2} \mathcal{O}_\mathcal{S}(N_{a_1} + N_{a_2}) \twoheadrightarrow \bigoplus_{l_1 < l_2} \mathcal{O}_\mathcal{S}(M_{l_1} + M_{l_2}) \longrightarrow 0, \\
0 &\longrightarrow \underaccent{\tilde}{\mathcal{V} \otimes \bigoplus_{l=1}^{n_p} \mathcal{O}_\mathcal{S}(M_l)} \hookrightarrow \bigoplus_{a,l} \mathcal{O}_\mathcal{S}(N_a + M_l) \twoheadrightarrow \bigoplus_{l_1,l_2} \mathcal{O}_\mathcal{S}(M_{l_1} + M_{l_2}) \longrightarrow 0,
\end{aligned}
\tag{7.70}
$$

where we made use of the fact that for line bundles $\mathcal{L}_k$ the following holds:

$$
\mathcal{O}_\mathcal{S}(\mathcal{L}_1) \otimes \mathcal{O}_\mathcal{S}(\mathcal{L}_2) = \mathcal{O}_\mathcal{S}(\mathcal{L}_1 + \mathcal{L}_2) \quad \text{and}
\tag{7.71}
$$

$$
\Lambda^2 \left( \bigoplus_{k=1}^m \mathcal{O}_\mathcal{S}(\mathcal{L}_k) \right) = \bigoplus_{k_1 < k_2} \mathcal{O}_\mathcal{S}(\mathcal{L}_{k_1} + \mathcal{L}_{k_1}).
\tag{7.72}
$$

In (7.70) we have furthermore introduced the curly underline in order to indicate that this is the bundle whose cohomology we want to determine via this specific sequence. Since from here many sequences will arise, this way we want to keep track of which bundle cohomology groups we already determined and which not.

Now (7.70) provides everything to obtain the cohomology of $\Lambda^2 \mathcal{V}$. Everything we have to do is to derive the long exact sequences of the latter two short exact sequences, then determine the cohomology of $Q_2$ as well as the tensor product that appears in the last sequence and plug the results into the long exact sequence corresponding to the first sequence in (7.70). The maps in these sequences are canonically constructed from the maps $F_a{}^l$ in the monad (7.43).

Non-exact monad

For the non-exact monad the calculation works similar but we will have to do the procedure described in last paragraph twice, since we do have two exact sequences (7.65) from the one non-exact complex (7.61). Applying table 7.6 to the second sequence of (7.65) we find

$$0 \longrightarrow \Lambda^2\mathcal{V} \hookrightarrow Q_1 \twoheadrightarrow \mathcal{V} \otimes \bigoplus_{l=1}^{n_p} \mathcal{O}_S(M_l) \longrightarrow 0\,,$$

$$0 \longrightarrow Q_1 \hookrightarrow \Lambda^2 \mathcal{E}_S \twoheadrightarrow \bigoplus_{l_1 < l_2} \mathcal{O}_S(M_{l_1} + M_{l_2}) \longrightarrow 0\,,$$

$$0 \longrightarrow \mathcal{V} \otimes \bigoplus_{l=1}^{n_p} \mathcal{O}_S(M_l) \hookrightarrow \bigoplus_{a,l} \mathcal{O}_S(N_a + M_l) \twoheadrightarrow \bigoplus_{l_1,l_2} \mathcal{O}_S(M_{l_1} + M_{l_2}) \longrightarrow 0\,,$$

(7.73)

and in order to get the auxiliary sheaf $Q_1$ we need to figure out what $\Lambda^2 \mathcal{E}_S$ is. This we can do by applying table 7.6 to the first sequence of (7.65) resulting in the following three sequences to determine

$$0 \longrightarrow \bigoplus_{a=1}^{n_\Lambda} \mathcal{O}_S(N_a) \otimes \mathcal{E}_S \hookrightarrow Q_2 \twoheadrightarrow \Lambda^2 \mathcal{E}_S \longrightarrow 0\,,$$

$$0 \longrightarrow \bigoplus_{a_1 < a_2} \mathcal{O}_S(N_{a_1} + N_{a_2}) \hookrightarrow \bigoplus_{l_1 < l_2} \mathcal{O}_S(M_{l_1} + M_{l_2}) \twoheadrightarrow Q_2 \longrightarrow 0\,,$$

$$0 \longrightarrow \bigoplus_{i,a} \mathcal{O}_S(L_i + N_a) \hookrightarrow \bigoplus_{a_1,a_2} \mathcal{O}_S(N_{a_1} + N_{a_2}) \twoheadrightarrow \bigoplus_{a=1}^{n_\Lambda} \mathcal{O}_S(N_a) \otimes \mathcal{E}_S \longrightarrow 0\,.$$

(7.74)

Also here the mappings can be derived canonically from mappings $F_a{}^l$ as well as $E_i{}^a$ in the original monad (7.61) by keeping track of them while performing the tensor products. Obviously we are faced with a lot of sequences for the determination of $\Lambda^2 \mathcal{V}$ since not only the six sequences in (7.73) and (7.74) appear but one additional for each hypersurface and line bundle in these sequences will

## 7.4. Vector bundle-valued cohomology

also arise due to the Koszul sequence (6.31). Also these calculations can be done algorithmically and as long as the mappings do not have to be computed explicitly our program **cohomCalg Koszul** extension will also provide the computations.

### 7.4.3 Cohomology of the endomorphism bundle of $\mathcal{V}$

Lets us now come to our last bundle whose cohomology we would like to determine. This is the one which involves most the steps and sequences. In fact even for simple examples it is a pain to perform the calculations by hand and we highly recommend to use our program **cohomCalg Koszul** extension. Even though it will not be able to perform the computation for all possible scenarios yet, it will provide a big help and in fact in most the examples one only hast to compute one specific map by hand and the rest is done by the computer.

In heterotic compactifications one often needs to count the number of bundle deformations of a bundle that lives on a Calabi-Yau manifold given as a submanifold of a toric variety. These are counted by the dimension of endomorphism bundle-valued cohomology groups. Specifically we need the dimension of

$$H^1\left(S;\operatorname{End}(\mathcal{V})\right) =: H^1_{\mathcal{S}}\left(\operatorname{End}(\mathcal{V})\right). \tag{7.75}$$

It is possible to relate this to the computation of cohomology groups with values in the tensor product of the bundle with its dual, i.e.

$$h^1_{\mathcal{S}}\left(\operatorname{End}(\mathcal{V})\right) = h^1_{\mathcal{S}}\left(\mathcal{V} \otimes \mathcal{V}^*\right). \tag{7.76}$$

To obtain a formula for this kind of bundle we can simply employ the corresponding monad sequence (7.43) or (7.61). Since we are usually calculating only the dimension of the cohomology, it comes in handy to put constraints to some dimensions that will help to obtain a unique result purely from exactness considerations and ignoring the actual mappings. Namely, if we know that the bundle is actually stable, we can put

$$h^0_{\mathcal{S}}(\mathcal{V} \otimes \mathcal{V}^*) = h^{\dim(\mathcal{S})}_{\mathcal{S}}(\mathcal{V} \otimes \mathcal{V}^*) = 1 \tag{7.77}$$

and in addition one can use that

$$h^1_{\mathcal{S}}(\mathcal{V} \otimes \mathcal{V}^*) = h^{\dim(\mathcal{S})-1}_{\mathcal{S}}(\mathcal{V} \otimes \mathcal{V}^*). \tag{7.78}$$

## Exact monad

As before let us start with the simpler case of the exact monad (7.43). To obtain the cohomology of $\mathcal{V} \otimes \mathcal{V}^*$ we have to take the tensor product of the dual exact monad (7.47) with $\mathcal{V}$ from the left. This will then give us two new bundles whose cohomology we have to determine, namely

$$\mathcal{V} \otimes \bigoplus_{a=1}^{n_\Lambda} \mathcal{O}_S(-N_a) \quad \text{and} \quad \mathcal{V} \otimes \bigoplus_{l=1}^{n_p} \mathcal{O}_S(-M_l). \tag{7.79}$$

To obtain these we can simply tensor (7.43) separately with each of the direct sums of line bundles as in (7.79). In total we will therefore be left with the following three sequences:

$$0 \longrightarrow \mathcal{V} \otimes \bigoplus_{l=1}^{n_p} \mathcal{O}_S(-M_l) \hookrightarrow \mathcal{V} \otimes \bigoplus_{a=1}^{n_\Lambda} \mathcal{O}_S(-N_a) \twoheadrightarrow \underline{\mathcal{V} \otimes \mathcal{V}^*} \longrightarrow 0,$$

$$0 \longrightarrow \underline{\mathcal{V} \otimes \bigoplus_{l=1}^{n_p} \mathcal{O}_S(-M_l)} \hookrightarrow \bigoplus_{a,l} \mathcal{O}_S(N_a - M_l) \twoheadrightarrow \bigoplus_{l_1,l_2} \mathcal{O}_S(M_{l_1} - M_{l_2}) \longrightarrow 0,$$

$$0 \longrightarrow \underline{\mathcal{V} \otimes \bigoplus_{a=1}^{n_\Lambda} \mathcal{O}_S(-N_a)} \hookrightarrow \bigoplus_{a_1,a_2} \mathcal{O}_S(N_{a_1} - N_{a_2}) \twoheadrightarrow \bigoplus_{l,a} \mathcal{O}_S(M_l - N_a) \longrightarrow 0, \tag{7.80}$$

with the maps induced by the maps in the sequences (7.43) and (7.47).

## Non-exact monad

Tensoring the dual sequence to the second sequence of (7.65) with $\mathcal{V}$ will give

$$0 \longrightarrow \mathcal{V} \otimes \bigoplus_{l=1}^{n_p} \mathcal{O}_S(-M_l) \hookrightarrow \mathcal{V} \otimes \mathcal{E}_S^* \twoheadrightarrow \underline{\mathcal{V} \otimes \mathcal{V}^*} \longrightarrow 0. \tag{7.81}$$

and tensoring the dualized first sequence of (7.65) with $\mathcal{V}$ results in

$$0 \longrightarrow \underline{\mathcal{V} \otimes \mathcal{E}_S^*} \hookrightarrow \mathcal{V} \otimes \bigoplus_{a=1}^{n_\Lambda} \mathcal{O}_S(-N_a) \longrightarrow \mathcal{V} \otimes \bigoplus_{i=1}^{n_F} \mathcal{O}_S(-L_i) \longrightarrow 0, \tag{7.82}$$

So what is left to do is do resolve the remaining missing bundles

$$\mathcal{V} \otimes \bigoplus_{l=1}^{n_p} \mathcal{O}_S(-M_l), \quad \mathcal{V} \otimes \bigoplus_{a=1}^{n_\Lambda} \mathcal{O}_S(-N_a), \quad \mathcal{V} \otimes \bigoplus_{i=1}^{n_F} \mathcal{O}_S(-L_i). \tag{7.83}$$

## 7.4. Vector bundle-valued cohomology

For each of these products we will need another two sequences that arise by tensoring the corresponding sums of line bundles with both sequences in (7.65):

$$0 \longrightarrow \bigoplus_{i,l} \mathcal{O}_\mathcal{S}(L_i - M_l) \hookrightarrow \bigoplus_{a,l} \mathcal{O}_\mathcal{S}(N_a - M_l) \longrightarrow \mathcal{E}_\mathcal{S} \otimes \underbrace{\bigoplus_{l=1}^{n_p} \mathcal{O}_\mathcal{S}(-M_l)}_{} \longrightarrow 0 \,,$$

$$0 \longrightarrow \mathcal{V} \otimes \underbrace{\bigoplus_{l=1}^{n_p} \mathcal{O}_\mathcal{S}(-M_l)}_{} \hookrightarrow \mathcal{E}_\mathcal{S} \otimes \bigoplus_{l=1}^{n_p} \mathcal{O}_\mathcal{S}(-M_l) \longrightarrow \bigoplus_{l_1,l_2} \mathcal{O}_\mathcal{S}(M_{l_1} - M_{l_2}) \longrightarrow 0 \,.$$

(7.84)

$$0 \longrightarrow \bigoplus_{i,a} \mathcal{O}_\mathcal{S}(L_i - N_a) \hookrightarrow \bigoplus_{a_1,a_2} \mathcal{O}_\mathcal{S}(N_{a_1} - N_{a_2}) \longrightarrow \mathcal{E}_\mathcal{S} \otimes \underbrace{\bigoplus_{a=1}^{n_\Lambda} \mathcal{O}_\mathcal{S}(-N_a)}_{} \longrightarrow 0 \,,$$

$$0 \longrightarrow \mathcal{V} \otimes \underbrace{\bigoplus_{a=1}^{n_\Lambda} \mathcal{O}_\mathcal{S}(-N_a)}_{} \hookrightarrow \mathcal{E}_\mathcal{S} \otimes \bigoplus_{a=1}^{n_\Lambda} \mathcal{O}_\mathcal{S}(-N_a) \longrightarrow \bigoplus_{l,a} \mathcal{O}_\mathcal{S}(M_l - N_a) \longrightarrow 0 \,.$$

(7.85)

$$0 \longrightarrow \bigoplus_{i_1,i_2} \mathcal{O}_\mathcal{S}(L_{i_1} - L_{i_2}) \hookrightarrow \bigoplus_{a,i} \mathcal{O}_\mathcal{S}(N_a - L_i) \longrightarrow \mathcal{E}_\mathcal{S} \otimes \bigoplus_{i=1}^{n_F} \mathcal{O}_\mathcal{S}(-M_l) \longrightarrow 0 \,,$$

$$0 \longrightarrow \mathcal{V} \otimes \underbrace{\bigoplus_{i=1}^{n_F} \mathcal{O}_\mathcal{S}(-L_i)}_{} \hookrightarrow \mathcal{E}_\mathcal{S} \otimes \bigoplus_{i=1}^{n_F} \mathcal{O}_\mathcal{S}(-L_i) \longrightarrow \bigoplus_{l,i} \mathcal{O}_\mathcal{S}(M_l - L_i) \longrightarrow 0 \,.$$

(7.86)

# Chapter 8

# Purely Combinatorial Approach to Cohomology

Up to now we have seen various methods to obtain the cohomology of topological spaces such as manifolds or vector bundles. In all cases, the task eventually boiled down to calculating the cohomology of some complex, of cochains for example. We have also seen besides that there was always the combinatorial task of counting Laurent monomials and in many cases this was even the main part since we could obtain the cohomology of the corresponding complex just by making use of exactness and without the explicit calculation of the maps. In fact purely combinatorial approaches to calculate cohomologies via lattice polytopes already exist. The first description of the Hodge numbers of a Calabi-Yau hypersurface in this fashion due to Batyrev [76] was followed by the generalization to complete intersections in higher-dimensional ambient spaces by Batyrev and Borisov [107]. Recently, there were also attempts to calculate bundle deformations of the tangent bundle of a Calabi-Yau three-fold in such a way [108,109]. In this chapter we want to show ways to relate the ingredients of such calculations to cohomology groups of line bundles of the corresponding ambient space which may allow a deeper insight to the combinatorial formulæ.

## 8.1 Lattice polytopes and Calabi-Yau hypersurfaces

As explained in detail in chapter 3, the geometry of a toric variety can be described by its fan which itself is defined as a triangulation of a given reflexive polytope, where lattice polytope is called *reflexive*, if its polar polytope is a lattice polytope, as well. Let $\Delta^\circ$ be such a polytope, then its *polar polytope* $\Delta$ is defined as

$$\Delta := \left\{ m \in \mathbb{Z}^d : \langle n, m \rangle \geq -1 \; \forall n \in \Delta^\circ \right\}. \tag{8.1}$$

While the vertices of the reflexive polytope $\Delta°$ of a given toric variety $\mathbb{P}_\Sigma$ represent the homogeneous coordinates as well as the equivalence relations between them, the lattice points of the polar polytope $\Delta$ represent a Calabi-Yau hypersurface $\mathcal{M}$ in $\mathbb{P}_\Sigma$. This hypersurface is defined by an equation $G = 0$, where $G$ is a sum of monomials induced from the lattice points $m$ of $\Delta$. The term corresponding to a fixed $m$ is then proportional to

$$\prod_{\rho \in \Delta°} z_\rho^{\langle m, \rho \rangle + 1}. \tag{8.2}$$

The polytope $\Delta$ is also called the *Newton polytope* of $\mathcal{M}$. On the other hand we can consider the vertices of $\Delta$ to be the defining data of some other toric variety and its polar polytope $\Delta°$ to be the Calabi-Yau hypersurface of $\Delta$. In this way we are able to relate two apparently completely different Calabi-Yau varieties with each other. For such Calabi-Yau hypersurfaces this relation is called *mirror symmetry*, see [110] for an exhaustive treatment. Since $G$ is made out of terms from (8.2) with different coefficients and due to the fact that $G$ describes the Calabi-Yau manifold, it is clear that the complex structure deformations are very much depending on $\Delta$. On the other hand, vertices in the polytope $\Delta°$ correspond to homogeneous coordinates $x_k$ and can therefore define the toric divisors

$$\mathcal{D}_k := \{x \in \mathbb{P}_\Sigma \mid x_k = 0\} \tag{8.3}$$

which will also define divisors on the Calabi-Yau. Hence it seems reasonable that $\Delta°$ strongly influences the number of Kähler moduli.

The mirror symmetry conjecture, while it follows trivially from conformal field theory point of view, is geometrically highly non-trivial and states that for every Calabi-Yau threefold $\mathcal{M}$ with Hodge numbers $(h^{2,1}(\mathcal{M}), h^{1,1}(\mathcal{M}))$ there exists a mirror three-fold $\hat{\mathcal{M}}$ with exchanged Hodge numbers $h^{2,1}(\hat{\mathcal{M}}) = h^{1,1}(\mathcal{M})$ and $h^{1,1}(\hat{\mathcal{M}}) = h^{2,1}(\mathcal{M})$. Thus, mirror symmetry exchanges complex structure deformations of $\mathcal{M}$ with the Kähler deformations of $\hat{\mathcal{M}}$ and vice versa. Mirror symmetry for hypersurfacses in $d$-dimensional toric varieties has the precise mathematical meaning of a simple exchange of the role of the two polytopes $\Delta$ and $\Delta°$. This is reflected in a nice way in the well-known combinatorial formula for the Hodge numbers of a Calabi-Yau hypersurface, first derived by Batytev [76]:

$$h^{1,1}(\mathcal{M}) = l(\Delta°) - d - 1 - \sum_{\dim(y)=0} l^*(y^\vee) + \sum_{\dim(y)=1} l^*(y) \cdot l^*(y^\vee), \tag{8.4}$$

$$h^{d-2,1}(\mathcal{M}) = l(\Delta) - d - 1 - \sum_{\dim(y_\circ)=0} l^*(y_\circ^\vee) + \sum_{\dim(y_\circ)=1} l^*(y_\circ) \cdot l(y_\circ^\vee). \tag{8.5}$$

## 8.1. Lattice polytopes and Calabi-Yau hypersurfaces

Here $y$ and $y_\circ$ are faces of $\Delta$ and $\Delta^\circ$, respectively, $l$ is the number of points of a face and $l^*$ is the number of interior points of a face. The dual face of an $r$ dimensional face $y \subset \Delta$ is a $(d - r - 1)$-dimensional face $y^\vee \subset \Delta^\circ$ defined by

$$y^\vee := \{n \in \Delta^\circ : \langle n, m \rangle = -1 \ \forall \ m \in y\}. \tag{8.6}$$

From the two equations (8.4) and (8.5) it is clear that the Calabi-Yau hypersurfaces $\mathcal{M}_\Delta$ and $\mathcal{M}_{\Delta^\circ}$ associated to the polytopes $\Delta$ and $\Delta^\circ$ have mirror symmetric Hodge numbers. Formula (8.4) counts the Kähler deformations of a given Calabi-Yau variety while the complex structure deformations are counted by equation (8.5). The first terms in those equations correspond to toric and polynomial deformations while the last term, where faces and dual faces get mixed up, corresponds to non-toric, i.e. non-polynomial deformations, respectively.

### 8.1.1 The Batyrev formulæ in terms of line bundle cohomology

As we have seen explicitly in subsection 7.3.1, it is also possible to obtain the Hodge numbers by making use of the Euler sequence (7.7) and the Koszul complex (6.31). Therefore the contributions to equations (8.4) and (8.5) have their origin in the cohomology of line bundles on the ambient space. The observation for three dimensional hypersurfaces is that all contributions to the polynomial deformations in the Batyrev formula arise from global sections of line bundles on the ambient space, namely from $h^0_{\mathbb{P}_\Sigma}(\cdot)$ for some divisor. In contrast, the non-polynomial deformations of the complex structure, which can also be associated with the twisted sector of the corresponding Gepner model, arise as $h^1_{\mathbb{P}_\Sigma}(\cdot)$ contributions.

In summary, we have observed the following identification of the various combinatorial contributions in the Batyrev formula with line bundle cohomologies:

$$\begin{aligned} \mathfrak{h}^{2,1}_0(\mathcal{M}) &= l(\Delta) - d - 1 - \sum_{\dim(y_\circ)=0} l^*(y_\circ^\vee), \\ \mathfrak{h}^{2,1}_1(\mathcal{M}) &= \sum_{\dim(y_\circ)=1} l^*(y_\circ) \cdot l(y_\circ^\vee), \end{aligned} \tag{8.7}$$

where we used the notation $\mathfrak{h}^{p,q}_i$ introduced in (7.28). For the hypersurface case we could not find any example where the second or any higher cohomology contributed to this particular Hodge number. This seems to fit the intuition of cohomology, since $h^1$ contributions represent cochains that are basically sections, defined on patches with dimension $\dim(\mathbb{P}_\Sigma - 1)$. So they can still be global sections on the subvariety if the subvariety is of codimension one. Similarly we would expect contributions $\mathfrak{h}_i^{p,q}$ for higher $i$ if we consider subvarieties of higher dimension.

For a K3 surface embedded in $\mathbb{P}^3$ the situation is a little different. Since it has only dimension 2, the tangent bundle cohomology ends of course with $h_{\mathcal{M}}^2(T_\mathcal{M})$. Usually the $h_{\mathbb{P}_\Sigma}^0(\cdot)$ and $h_{\mathbb{P}_\Sigma}^1(\cdot)$ contributions "flow" in the complex structure deformations and the higher ones like $h_{\mathbb{P}_\Sigma}^{n-1}(\cdot)$ and $h_{\mathbb{P}_\Sigma}^n(\cdot)$ to the Kähler deformations. Since the K3 has only a single non-trivial Hodge number that is not on the edge of the Hodge diamond, both sides contribute to it and we can see a split of this Hodge number $h_{K3}^{1,1} = 20$ into $19 + 1$. We also know that for this K3 there is a split of the 20 deformations into 19 algebraic ones which would correspond to those coming from the $h_{\mathbb{P}_\Sigma}^0(\cdot)$ and 1 non-algebraic one associated to the $h_{\mathbb{P}_\Sigma}^2(\cdot)$ contributions.

One can easily check this with the cohomCalg Koszul extension [69] by using the "Verbose5" option and following the contributions through the long exact sequences.

## 8.2 Cayley polytopes and CICYs

The next step is to extend the above formula to a formula for a complete intersection Calabi-Yau (CICY[1]) in a five-dimensional toric variety. This was done by Batyrev and Borisov shortly after the hypersurface case [107]. We will now describe how to do that, following [111].

### 8.2.1 Nef partitions and their Cayley polytopes

As before we start with a reflexive polytope $\Delta^\circ$, which is this time describing a five-dimensional toric variety $\mathbb{P}_\Sigma$. Consider now a partition of all vertices in $\Delta^\circ$ into disjoint subsets $V_i$ and define the polytopes following from those vertices as

$$\Delta_i^\circ := \mathrm{Conv}(V_i, 0) \quad \text{for } i = 1, \ldots, c, \tag{8.8}$$

where $\mathrm{Conv}(\cdot)$ denotes the convex hull of a vertex set. Such a partition is called a nef (numerically effective) partition, if the Minkowski sum of all $\Delta_i^\circ$ forms a

---

[1] Some authors refer to CICY when they mean the complete intersection Calabi-Yaus inside products of projective spaces. I do not understand why they use this notation in particular since the name does not suggest any relation to such a particular ambient space and since there is not really a different shorthand for complete intersections in arbitrary ambient spaces we will still use this one and hope not to confuse the reader with it.

## 8.2. Cayley polytopes and CICYs

reflexive polytope. As a reminder, the *Minkowski sum* of two vertex sets is defined by

$$\Delta_1^\circ + \Delta_2^\circ := \text{Conv}(\{v + v' : v \in \Delta_1^\circ, \ v' \in \Delta_2^\circ\}). \tag{8.9}$$

It is easy to see that one can associate a complete intersection Calabi-Yau variety to such a partition. It is the intersection of $c$ hypersurfaces, where each hypersurface $\mathcal{S}_i$ corresponds to a sum of divisors $\mathcal{D}_{i,k} = \{x_k = 0\}$. Here each $x_k$ is the homogeneous coordinate that belongs to the vertex $v_k \in \Delta_i^\circ$:

$$\mathcal{S}_i = \sum_{k \in K} \mathcal{D}_{i,k}, \text{ where } K = \{k \in \mathbb{N} \text{ such that } v_k \in \Delta_i^\circ\}. \tag{8.10}$$

So one might assume that one treats those "subpolytopes" in the same fashion as in the hypersurface case, but this is not quite right. Instead of dealing with the polytope that describes the toric ambient space, it is necessary to construct a different kind of polytope that respects the nef partition explicitly. This polytope is called the *Cayley polytope* and is defined by

$$P^* := \text{Conv}(\Delta_1^\circ \times e_1, \ldots, \Delta_c^\circ \times e_c). \tag{8.11}$$

Here $e_1, \ldots, e_c$ is the canonical basis of $\mathbb{Z}^c$ and hence the Cayley polytope is a $d + c$ dimensional polytope, requiring a somewhat more sophisticated treatment compared to the aforementioned cases. For instance, it is not possible to compute a well-defined polar polytope anymore, and we need a different way to find a dual polytope for the Cayley polytope. It goes as follows: We consider the cone that supports the Cayley polytope, called the Cayley cone $C^*$, and calculate its dual cone $C$ via

$$C := \{n \in \mathbb{R}^d \times \mathbb{R}^c : \langle m, n \rangle \geq 0, \ \forall m \in C^*\}. \tag{8.12}$$

One can see that the dual Cayley cone $C$ also supports a polytope called the dual Cayley polytope. If we take slices of this one by successively putting the $i^{\text{th}}$ coordinate in $\{x_{d+1}, \ldots, x_{d+i}, \ldots, x_{d+c}\}$ to one and all the others in this set to zero for $i = 1, \ldots, c$, we get polytopes $\nabla_i$ whose Minkowski sum corresponds to the polar polytope $\Delta$ of our original polytope $\Delta^\circ$, namely

$$\Delta = \nabla_1 + \cdots + \nabla_c. \tag{8.13}$$

On the other hand, the convex hull $\nabla = \text{Conv}(\{\nabla_i, \ i = 1, \ldots, c\})$ also represents a reflexive polytope that is polar to the Minkowski sum of the dual nef partition, i.e. $\nabla^\circ = \Delta_1^\circ + \ldots + \Delta_c^\circ$ which gives rise to a complete intersection Calabi-Yau in the toric variety corresponding to this polytope. In summary, we therefore have the following relation between $\Delta^\circ$ and $\nabla$:

$$\Delta^\circ = \operatorname{Conv}(\{\Delta_i^\circ,\ i=1,\ldots,c\}), \quad \nabla = \operatorname{Conv}(\{\nabla_i,\ i=1,\ldots,c\}),$$
$$\Delta = \nabla_1 + \ldots + \nabla_c, \quad \nabla^\circ = \Delta_1^\circ + \ldots + \Delta_c^\circ, \quad (8.14)$$
$$P^*_{\Delta^\circ} = P_\nabla, \quad P^*_\nabla = P_{\Delta^\circ}.$$

So we need two different polytopes in order to find the two dual complete intersection Calabi-Yau manifolds. One of them is the complete intersection of hypersurfaces corresponding to vertex sets $\Delta_i^\circ$ that intersect in the toric variety built from $\Delta^\circ$, the other one is a complete intersection of hypersurfaces corresponding to vertex sets $\nabla_i$ that intersect in the toric variety built from $\nabla$. Clearly for the hypersurface case we find that $\nabla = \Delta$ and $\nabla^\circ = \Delta^\circ$ which reduces everything to the setting of section 8.1.

### 8.2.2 Example: $\mathbb{P}^5_{112233}$

| Vertices of $\Delta^\circ$ | Vertices of $\Delta$ |
|---|---|
| $v_1 = (\,-1,\ -2,\ -2,\ -3,\ -3\,)$ | $w_1 = (\,-1,\ -1,\ -1,\ -1,\ -1\,)$ |
| $v_2 = (\ \ \ 1,\ \ \ 0,\ \ \ 0,\ \ \ 0,\ \ \ 0\,)$ | $w_2 = (\ \ \ 3,\ -1,\ -1,\ -1,\ -1\,)$ |
| $v_3 = (\ \ \ 0,\ \ \ 1,\ \ \ 0,\ \ \ 0,\ \ \ 0\,)$ | $w_3 = (\,-1,\ \ \ 3,\ -1,\ -1,\ -1\,)$ |
| $v_4 = (\ \ \ 0,\ \ \ 0,\ \ \ 1,\ \ \ 0,\ \ \ 0\,)$ | $w_4 = (\,-1,\ -1,\ \ \ 5,\ -1,\ -1\,)$ |
| $v_5 = (\ \ \ 0,\ \ \ 0,\ \ \ 0,\ \ \ 1,\ \ \ 0\,)$ | $w_5 = (\,-1,\ -1,\ -1,\ \ \ 5,\ -1\,)$ |
| $v_6 = (\ \ \ 0,\ \ \ 0,\ \ \ 0,\ \ \ 0,\ \ \ 1\,)$ | $w_6 = (\,-1,\ -1,\ -1,\ -1,\ \ 11\,)$ |
| (a) Lattice polytope of $\mathbb{P}^5_{112233}$ | (b) Polar polytope |

**Table 8.1.:** Lattice polytope $\Delta^\circ$ for the weighted projective space $\mathbb{P}^5_{112233}$ and its polar polytope $\Delta$.

Let us consider a specific example to illustrate what we learned so far. Let $\mathbb{P}_\Sigma$ be the five-dimensional weighted projective space $\mathbb{P}^5_{112233}$, which is described by the polytope $\Delta^\circ_{\mathbb{P}^5_{112233}}$ from table 8.1a. Consider a partition of $\Delta^\circ$ into the subsets $\Delta_1^\circ = \operatorname{Conv}(\{v_1, v_2, v_4\}, 0)$ and $\Delta_2^\circ = \operatorname{Conv}(\{v_3, v_5, v_6\}, 0)$. One can show that $\Delta_1^\circ + \Delta_2^\circ$ is a reflexive polytope and hence this partition is indeed a nef partition. Using (8.11) we can calculate $P^*_{\Delta^\circ}$, see table 8.2a, and by employing (8.14) the dual Cayley polytope $P_{\Delta^\circ}$ given in table 8.2b follows. From $P_{\Delta^\circ}$ it is quite easy to read off $\nabla$ and its nef partition $\{\nabla_1, \nabla_2\}$ and we can confirm that the Minkowski sum

$$\nabla_1 + \nabla_2 \quad (8.15)$$

## 8.2. Cayley polytopes and CICYs

of this partition indeed equals the polar polytope of $\Delta^\circ$ in table 8.1b.

| Vertices of $P^*_{\Delta^\circ}$ |
|---|
| $\tilde{v}_1 = (\ -1,\ -2,\ -2,\ -3,\ -3,\ \ 1,\ \ 0\ )$ |
| $\tilde{v}_2 = (\ \ \ 1,\ \ \ 0,\ \ \ 0,\ \ \ 0,\ \ \ 0,\ \ 1,\ \ 0\ )$ |
| $\tilde{v}_3 = (\ \ \ 0,\ \ \ 1,\ \ \ 0,\ \ \ 0,\ \ \ 0,\ \ 0,\ \ 1\ )$ |
| $\tilde{v}_4 = (\ \ \ 0,\ \ \ 0,\ \ \ 1,\ \ \ 0,\ \ \ 0,\ \ 1,\ \ 0\ )$ |
| $\tilde{v}_5 = (\ \ \ 0,\ \ \ 0,\ \ \ 0,\ \ \ 1,\ \ \ 0,\ \ 0,\ \ 1\ )$ |
| $\tilde{v}_6 = (\ \ \ 0,\ \ \ 0,\ \ \ 0,\ \ \ 0,\ \ \ 1,\ \ 0,\ \ 1\ )$ |
| $\tilde{v}_7 = (\ \ \ 0,\ \ \ 0,\ \ \ 0,\ \ \ 0,\ \ \ 0,\ \ 1,\ \ 0\ )$ |
| $\tilde{v}_8 = (\ \ \ 0,\ \ \ 0,\ \ \ 0,\ \ \ 0,\ \ \ 0,\ \ 0,\ \ 1\ )$ |

(a) The Cayley polytope

| Vertices of $P_{\Delta^\circ}$ |
|---|
| $\tilde{w}_1 = (\ \ \ 0,\ -1,\ \ \ 0,\ -1,\ -1,\ \ 1,\ \ 0\ )$ |
| $\tilde{w}_2 = (\ \ \ 2,\ -1,\ \ \ 0,\ -1,\ -1,\ \ 1,\ \ 0\ )$ |
| $\tilde{w}_3 = (\ \ \ 0,\ \ \ 1,\ \ \ 0,\ -1,\ -1,\ \ 1,\ \ 0\ )$ |
| $\tilde{w}_4 = (\ \ \ 0,\ -1,\ \ \ 3,\ -1,\ -1,\ \ 1,\ \ 0\ )$ |
| $\tilde{w}_5 = (\ \ \ 0,\ -1,\ \ \ 0,\ \ \ 2,\ -1,\ \ 1,\ \ 0\ )$ |
| $\tilde{w}_6 = (\ \ \ 0,\ -1,\ \ \ 0,\ -1,\ \ \ 5,\ \ 1,\ \ 0\ )$ |
| $\tilde{w}_7 = (\ -1,\ \ \ 0,\ -1,\ \ \ 0,\ \ \ 0,\ \ 0,\ \ 1\ )$ |
| $\tilde{w}_8 = (\ \ \ 1,\ \ \ 0,\ -1,\ \ \ 0,\ \ \ 0,\ \ 0,\ \ 1\ )$ |
| $\tilde{w}_9 = (\ -1,\ \ \ 2,\ -1,\ \ \ 0,\ \ \ 0,\ \ 0,\ \ 1\ )$ |
| $\tilde{w}_{10} = (\ -1,\ \ \ 0,\ \ \ 2,\ \ \ 0,\ \ \ 0,\ \ 0,\ \ 1\ )$ |
| $\tilde{w}_{11} = (\ -1,\ \ \ 0,\ -1,\ \ \ 3,\ \ \ 0,\ \ 0,\ \ 1\ )$ |
| $\tilde{w}_{12} = (\ -1,\ \ \ 0,\ -1,\ \ \ 0,\ \ \ 6,\ \ 0,\ \ 1\ )$ |

(b) The dual Cayley polytope

**Table 8.2.:** The Cayley polytope and the dual Cayley polytope corresponding to the nef partition $\Delta_1^\circ = \text{Conv}(\{v_1, v_2, v_4\}, 0)$ and $\Delta_2^\circ = \text{Conv}(\{v_3, v_5, v_6\}, 0)$ of $\mathbb{P}^5_{12233}$ as well as to the dual nef partition $\nabla_1$ and $\nabla_2$ coming from $\{\tilde{w}_1, \ldots, \tilde{w}_6\}$ and $\{\tilde{w}_7, \ldots, \tilde{w}_{12}\}$, respectively. The nef partition of the dual Cayley polytope can easily be read off from the last two columns.

### 8.2.3 The stringy $E$-function

The generalization of equations (8.4) and (8.5) to complete intersections will be a formula that counts faces in the Cayley and the dual Cayley polytope instead of the original one and its polar. The formula itself will also become a bit more complicated compared to the one for hypersurfaces.

In [107] Batyrev and Borisov introduced a generating function for the so-called stringy Hodge numbers of a CICY corresponding to the introduced Cayley cone above. These stringy Hodge numbers are equal to the usual Hodge numbers in case that a crepant resolution of the generically singular Calabi-Yau exists. They are given as coefficients of the stringy $E$-function, namely

$$E_{\text{st}}(\mathcal{S}; u, v) = \sum_{p,q} (-1)^{p+q} h_{\text{st}}^{p,q}(\mathcal{S})\, u^p v^q. \tag{8.16}$$

The generalization to arbitrary Gorenstein polytopes was done by Batyrev and Nill in [112]. There it was conjectured that the stringy $E$-function is actually a polynomial in $u,v$ which was proven recently by Nill and Schepers in [113]. In terms of the dual Cayley polytope, the stringy $E$-function can be expanded as

$$E(u,v) = \frac{1}{(uv)^c} \sum_{\emptyset \leq x \leq y \leq P} (-1)^{1+\dim x} u^{1+\dim y} S_x\left(\frac{u}{v}\right) S_{y^\vee}(uv) B_{[x,y]}(u^{-1}, v). \quad (8.17)$$

Here we sum over faces of $P$. These form an (Eulerian) partially ordered set (poset) where the partial ordering is given by

$$x \leq y \Leftrightarrow x \text{ is a face of } y \quad (8.18)$$

and denote the poset of all faces of $P$ by $\mathcal{P}$. Furthermore we define

$$\begin{aligned}[x,y] &:= \{z \in \mathcal{P} : x \leq z \leq y\} \quad \text{as well as} \\ y_C^\vee &:= \{n \in C : \langle m, n \rangle = 0, \ \forall m \in C^*\}\end{aligned} \quad (8.19)$$

to be the dual face of a face $y_C$ in $C$. Since there is a relation between faces in $C$ and $C^*$ we also get such a correspondence for faces in the supporting polytopes $P$ and $P^*$. The dual face to $y$ in $P$ is denoted by $y^\vee$ in $P^*$.

For some $k$-dimensional face $\mathcal{F}$, the polynomial $S_{\mathcal{F}}(t)$ is defined by

$$S_{\mathcal{F}}(t) := (1-t)^{k+1} \sum_{i=0}^{\infty} l(k\mathcal{F}) t^i. \quad (8.20)$$

For the last missing piece, let $\mathcal{P}'$ be some subposet of $\mathcal{P}$ with minimal element $\hat{0}$, maximal element $\hat{1}$ and $\dim(\hat{1}) = k - 1$. The Batyrev-Borisov polynomials $B_{\mathcal{P}'}(u,v)$ are defined recursively in terms of the polynomials $H_{\mathcal{P}'}(t)$ and $G_{\mathcal{P}'}(t)$. We set $H_{\mathcal{P}'}(t) = G_{\mathcal{P}'}(t) = 1$ if $k = 0$ and for $k > 0$ we define

$$\begin{aligned}H_{\mathcal{P}'}(t) &:= \sum_{\hat{0} < x < \hat{1}} (t-1)^{\dim(x)} G_{[x,\hat{1}]}(t), \\ G_{\mathcal{P}'}(t) &:= \tau_{<\frac{k}{2}}(1-t) H_{\mathcal{P}'}(t),\end{aligned} \quad (8.21)$$

where $\tau_{<\frac{k}{2}}$ is the truncation operator defined by its action on sums

$$\tau_{<\frac{k}{2}} \sum_{i=0}^{\infty} a_i t^i := \sum_{0 \leq i < \frac{k}{2}} a_i t^i. \quad (8.22)$$

The Batyrev-Borisov polynomials start also with $B_{\mathcal{P}'}(u,v) = 1$ for $k = 0$ and can

## 8.2. Cayley polytopes and CICYs

be read off from the following formula for $k > 0$:

$$\sum_{\hat{0}<x<\hat{1}} B_{[\hat{0},x]}(u,v) u^{k-\dim(x)+1} G_{[x,\hat{1}]}(u^{-1}v) := G_{\mathcal{P}'}(uv) . \tag{8.23}$$

### 8.2.4 Closed form expressions for $h^{1,1}$ and $h^{d-3,1}$ of a CICY

The above formula (8.17) might be elegant but is not at all easy to work with. Using some simplifying relations for the polynomials above, stated for instance in [107] or alternatively in [111] propositions 2.2 and 2.4, one can deduce the formulæ (8.4) and (8.5) from (8.17) (see Theorem 3.1 in [111]). Similarly in the same paper Doran and Novoseltsev deduced an explicit closed form expression for the Hodge numbers of Calabi-Yau varieties realized as a complete intersection of two hypersurfaces in a five-dimensional toric variety.

To present their result, we first recall that the nef partition is called indecomposable, if no subset of $\{\Delta_i^\circ, i = 1, \ldots, c\}$ exists whose Minkowski sum is reflexive, namely if we are not dealing with a product of Calabi-Yau manifolds. If this is the case, the combinatorial formula for the Hodge numbers is given as follows:

$$h^{1,1}(\mathcal{S}) = l(P^*) - 7$$
$$- \sum_{\dim(y)=0} l^*(2y^\vee) + \sum_{\dim(y)=1} l^*(y^\vee)$$
$$+ \sum_{\dim(y)=1} l^*(y) \cdot l^*(2y^\vee) - \sum_{\dim(y)=2} l^*(2y) \cdot l^*(y^\vee)$$
$$- \sum_{\substack{\dim(x)=2 \\ \dim(y)=3 \\ x<y}} l^*(x) \cdot l^*(y^\vee) + \sum_{\dim(y)=3} l^*(2y) \cdot l^*(y^\vee). \tag{8.24}$$

By taking all dual polytopes and faces we obtain of course the second Hodge number as

$$h^{2,1}(\mathcal{S}) = l(P) - 7 - \sum_{\dim(y_*)=0} l^*(2y_*^\vee) + \sum_{\dim(y_*)=1} l^*(y_*^\vee)$$
$$+ \sum_{\dim(y_*)=1} l^*(y_*) \cdot l^*(2y_*^\vee) - \sum_{\dim(y_*)=2} l^*(2y_*) \cdot l^*(y_*^\vee)$$
$$- \sum_{\substack{\dim(x_*)=2 \\ \dim(y_*)=3 \\ x_*<y_*}} l^*(x_*) \cdot l^*(y_*^\vee) + \sum_{\dim(y_*)=3} l^*(2y_*) \cdot l^*(y_*^\vee). \tag{8.25}$$

Here $x, y$ and $x_*, y_*$ are faces of $P$ and $P_*$, respectively. Having a close look at the proof of formulæ (8.24) and (8.25), one realizes that it is not too hard to generalize it to the case of two complete intersections in a six-dimensional ambient space, yielding a Calabi-Yau four-fold. Due to recent interest in such four-folds in the context of F-theory we deduced a closed form expression for such cases at least for the Hodge numbers $h^{1,1}$ and $h^{3,1}$ via the dual computation. The result is in the most generic form and no simplifications have been taken into account. For a Calabi-Yau four-fold $\mathcal{S}$ we find $h^{1,1}(\mathcal{S})$ along with its mirror dual $h^{3,1}(\mathcal{S})$:

$$\begin{aligned}h^{1,1}(\mathcal{S}) = l(P^*) - 8 &- \sum_{\dim(y)=0} [l^*(2y^\vee) - 7l^*(y^\vee)] \\&+ \sum_{\dim(y)=1} l^*(y^\vee) - \sum_{\dim(y)=2} [k(y) - 3] \cdot l^*(y^\vee) \\&+ \sum_{\dim(y)=1} l^*(y) \cdot [l^*(2y) - 6 \cdot l^*(y^\vee)] - \sum_{\substack{\dim(x)=1 \\ \dim(y)=2 \\ x<y}} l^*(x) \cdot l^*(y^\vee) \\&- \sum_{\substack{\dim(x)=2 \\ \dim(y)=3 \\ x<y}} l^*(x) \cdot l^*(y^\vee) + \sum_{\dim(y)=3} l^*(y^\vee) \cdot [l^*(2y) - 4l^*(y)] \,,\end{aligned} \quad (8.26)$$

$$\begin{aligned}h^{3,1}(\mathcal{S}) = l(P) - 8 &- \sum_{\dim(y_*)=0} [l^*(2y_*^\vee) - 7l^*(y_*^\vee)] \\&+ \sum_{\dim(y_*)=1} l^*(y_*^\vee) - \sum_{\dim(y_*)=2} [k(y_*) - 3] \cdot l^*(y_*^\vee) \\&+ \sum_{\dim(y_*)=1} l^*(y_*) \cdot [l^*(2y_*) - 6 \cdot l^*(y_*^\vee)] - \sum_{\substack{\dim(x_*)=1 \\ \dim(y_*)=2 \\ x_*<y_*}} l^*(x_*) \cdot l^*(y_*^\vee) \\&- \sum_{\substack{\dim(x_*)=2 \\ \dim(y_*)=3 \\ x_*<y_*}} l^*(x_*) \cdot l^*(y_*^\vee) + \sum_{\dim(y_*)=3} l^*(y_*^\vee) \cdot [l^*(2y_*) - 4l^*(y_*)] \,.\end{aligned} \quad (8.27)$$

Here $k(y)$ denotes the number of vertices of the face $y$. It is quite nice to have such an explicit formula at hand but one has also to admit that the actual calculation is not that easy to perform and may take some time. Our implementation in Macaulay2 using the `Polyhedra` package [114] needed at least ten minutes to produce the numbers. Using **cohomCalg** 9 [69] we were able to obtain the same results along with the remaining two Hodge numbers much faster for varieties where $h^{1,1}(\mathcal{S})$ is not too large.

## 8.2. Cayley polytopes and CICYs

$$\mathfrak{h}_5^{1,1}(\mathcal{S}) = l(P^*) - 7,$$

$$\mathfrak{h}_4^{1,1}(\mathcal{S}) = -\sum_{\dim(y)=0} l^*(2y^\vee) + \sum_{\dim(y)=1} l^*(y^\vee),$$

$$\mathfrak{h}_3^{1,1}(\mathcal{S}) = \sum_{\dim(y)=1} l^*(y) \cdot l^*(2y^\vee) - \sum_{\dim(y)=2} l^*(2y) \cdot l^*(y^\vee),$$

$$\mathfrak{h}_2^{1,1}(\mathcal{S}) = -\sum_{\substack{\dim(x)=2 \\ \dim(y)=3 \\ x<y}} l^*(x) \cdot l^*(y^\vee) + \sum_{\dim(y)=3} l^*(2y) \cdot l^*(y^\vee)$$

**Table 8.3.:** Relation between the terms in the combinatorial formula for $h^{1,1}(\mathcal{S})$ and line bundle cohomologies.

$$\mathfrak{h}_0^{2,1}(\mathcal{S}) = l(P) - 7 - \sum_{\dim(y_*)=0} l^*(2y_*^\vee) + \sum_{\dim(y_*)=1} l^*(y_*^\vee)$$

$$\mathfrak{h}_1^{2,1}(\mathcal{S}) = \sum_{\dim(y_*)=1} l^*(y_*) \cdot l^*(2y_*^\vee) - \sum_{\dim(y_*)=2} l^*(2y_*) \cdot l^*(y_*^\vee)$$

$$\mathfrak{h}_2^{2,1}(\mathcal{S}) = -\sum_{\substack{\dim(x_*)=2 \\ \dim(y_*)=3 \\ x_*<y_*}} l^*(x_*) \cdot l^*(y_*^\vee) + \sum_{\dim(y_*)=3} l^*(2y_*) \cdot l^*(y_*^\vee)$$

**Table 8.4.:** Relation between the terms in the combinatorial formula for $h^{2,1}(\mathcal{S})$ and line bundle cohomologies.

### 8.2.5 Correspondence to line bundle cohomologies of CICY

The question now is, whether also for these CICYs one can identify terms in the generalized Batyrev formulæ with line bundle cohomology classes of the ambient five-fold respectively six-fold.

Let us here restrict to the case of a CICY in an ambient five-fold, i.e. the formulæ (8.24) and (8.25). Indeed, we observed that, if we calculate those Hodge numbers via exact sequences and therefore via line bundle cohomologies of the ambient toric variety — using **cohomCalg** [69] as described in 7.3.2 — we find that the relation between the terms in the combinatorial formula for the $h^{1,1}(\mathcal{S})$ and the line bundle cohomologies applies according to table 8.3. Here $\mathfrak{h}_i^{p,q}(\mathcal{S})$ are defined as in section 7.3.2 equation (7.28). Similarly, we find the relations for $h^{2,1}(\mathcal{S})$ Hodge number as in table 8.4. For $h^{2,1}(\mathcal{S})$, as in (8.5), we note that terms where no mixing of faces and dual faces takes place correspond to global sections in line bundles whereas terms where such a mixing does happen correspond to higher cohomology classes of line bundles of $\mathbb{P}_\Sigma$. In (8.25) for one hypersurface we got at most $h^1_{\mathbb{P}_\Sigma}(\cdot)$ contributions, where now in case of two hypersurfaces we found at most $h^2_{\mathbb{P}_\Sigma}(\cdot)$ contributions to the complex structure moduli which meet what we expected.

Since the computations get quite involved, we have not yet identified the analogous correspondence for the case of a CICY in a toric six-fold, i.e. eq. (8.26) and (8.27), but have no doubt that it exists and takes a similar form.

# Chapter 9

# Target Space Dualities in the String Landscape

Dualities play a crucial role in string theory. The first investigation of superstring theories led to not only one but in fact to five different theories that all seemed to be independently consistent. To realize that these seemingly different frameworks are actually descriptions of the same thing, one has to use dualities that explain their relation. In particular the more generic framework of eleven-dimensional M-theory is connected to this web of dualities. Recently the twelve-dimensional theory, so-called F-theory, that can be considered as the non-perturbative limit of type IIB was studied extensively and especially the duality between heterotic string theories and F-theory arise geometrically in a very beautiful way.

The main focus of this chapter lies on a duality between different heterotic string compactifications that is referred to as *target space duality*. It was first observed and discussed in [94] and further in [95, 115] for heterotic string models whose phase space, which we introduced in chapter 2, contains a phase that is non-classically described by a Landau-Ginzburg orbifold. For such a configuration certain homogeneous polynomials were found to appear on equal footing. Since these polynomials have a geometric interpretation in the corresponding geometric phase of the same GLSM, i.e. they define the complex structure and the bundle structure, it is rather a matter of notation whether one describes one or the other. What appears to look trivial in the LG phase corresponds to a highly non-trivial duality of geometries. To be concrete, a given configuration $(\mathcal{M}, \mathcal{V})$ where $\mathcal{M}$ is a smooth Calabi-Yau three-fold and $\mathcal{V}$ is a stable holomorphic vector bundle will then have a finite number different equivalent descriptions

$$(\mathcal{M}, \mathcal{V}) \rightsquigarrow (\mathcal{M}_1, \mathcal{V}_1) \rightsquigarrow (\mathcal{M}_2, \mathcal{V}_2) \rightsquigarrow \ldots \rightsquigarrow (\mathcal{M}_m, \mathcal{V}_m). \quad (9.1)$$

Each duality in (9.1) corresponds to the exchange of a certain number of complex structure parameters with bundle parameters, and the number finite $m$ will be the maximal number of possibilities of a consistent exchange. Generalizations of these ideas were first suggested in [116, 117] where scenarios with singular dual configurations were considered. In order to resolve these singularities a blowup of the dual model was performed, increasing the number of Kähler parameters by one. It was shown there that in this specific example one could still find evidence for the fact that the smooth dual model is dual to the original by checking that the dimension of the full moduli space was still preserved even though in the dual model the Kähler moduli increased in number. So in some sense it is similar to a different kind of geometric duality referred to as *mirror symmetry* where an exchange of Kähler parameters and complex structure parameters takes place. The difference is though that for mirror symmetry it is only possible to find one mirror partner to a given manifold with precisely all of the complex structure parameters and Kähler parameters exchanged.

In the following we will investigate such target space dualities and further generalizations thereof. In particular we will give a concrete procedure for the construction of such models and will explicitly give examples for various cases where the dimension of the Kähler moduli space grows up to seven. Furthermore many scenarios are tested and described in full detail for models with vector bundles of rank three, four and five. Finally we will present a scan over more than 80,000 models and describe which cases provide evidence for the duality and in which cases the conjecture seems to fail. For the latter we also present an explanation for why they fail to meet the requirement of a duality.

## 9.1 The setup and assumptions

The starting point is a configuration for a heterotic model given by $(\mathcal{M}, \mathcal{V})$, where $\mathcal{M}$ is given as a complete intersection of $c$ hypersurfaces in a toric variety $\mathbb{P}_\Sigma$:

$$\mathcal{M} = \bigcap_{j=1}^{c} \mathcal{S}_j, \quad \mathcal{S}_j = \{G_j(x_i) = 0\} \subset \mathbb{P}_\Sigma =: \mathbb{P}_\Sigma[\mathcal{S}_1, ..., \mathcal{S}_c]. \tag{9.2}$$

The holomorphic vector bundles $\mathcal{V}$ we will consider are all coming from the following kind of monad:

$$0 \to \mathcal{O}_\mathcal{M}^{\oplus n_F} \xrightarrow{\otimes E^i_a} \bigoplus_{a=1}^{n_\Lambda} \mathcal{O}_\mathcal{M}(N_a) \xrightarrow{\otimes F_a{}^l} \bigoplus_{l=1}^{n_p} \mathcal{O}_\mathcal{M}(M_l) \to 0, \tag{9.3}$$

## 9.1. The setup and assumptions

where $\mathcal{V}$ is given by

$$\mathcal{V} = \frac{\ker(F)}{\operatorname{im}(E)}, \quad \operatorname{rk}(\mathcal{V}) = n_\Lambda - n_p - n_F. \tag{9.4}$$

The case where $n_F$ is zero is also allowed and corresponds to the case where the monad is exact

$$0 \to \mathcal{V} \longrightarrow \bigoplus_{a=1}^{n_\Lambda} \mathcal{O}_\mathcal{M}(N_a) \xrightarrow{\otimes F_a{}^l} \bigoplus_{l=1}^{n_p} \mathcal{O}_\mathcal{M}(M_l) \to 0, \tag{9.5}$$

and we have

$$\mathcal{V} = \ker(F), \quad \operatorname{rk}(\mathcal{V}) = n_\Lambda - n_p. \tag{9.6}$$

We always require the model $(\mathcal{M}, \mathcal{V})$ to have a proper smooth base $\mathcal{M}$ as well as a well defined smooth vector bundle $\mathcal{V}$. Furthermore the model we start with is always required to be physically sensible meaning that all conditions implied by anomaly cancellations etc. meaning that (2.41) is fully satisfied by $(\mathcal{M}, \mathcal{V})$.

To check whether the potentially dual model $(\widetilde{\mathcal{M}}, \widetilde{\mathcal{V}})$, whose construction we will illustrate in a second, is indeed a dual description of the initial configuration $(\mathcal{M}, \mathcal{V})$ a set of topological quantities should match. In our analysis we will always check for two specific conditions:

1. Matching spectrum: According to table 2.1 the following has to be satisfied:

$$H^1(\mathcal{M}; \Lambda^m \mathcal{V}) = H^1(\widetilde{\mathcal{M}}; \Lambda^m \widetilde{\mathcal{V}}), \quad \forall m = 1, ..., \operatorname{rk} \mathcal{V} - 1 \tag{9.7}$$

   for a rank $n$ bundle. We will in particular check this relation for $m = 1$ and $m = n-1$. In fact we can identify $H^1(\mathcal{M}; \Lambda^{\operatorname{rk}\mathcal{V}-1}\mathcal{V})$ with $H^{3-1}(\mathcal{M}; \Lambda^{\operatorname{rk}\mathcal{V}-(\operatorname{rk}\mathcal{V}-1)}\mathcal{V})$ and for stable bundles this corresponds in comparing

$$H^\bullet(\mathcal{M}; \mathcal{V}) = H^\bullet(\widetilde{\mathcal{M}}; \widetilde{\mathcal{V}}). \tag{9.8}$$

2. Matching moduli space dimension: In many cases this can be checked by counting the two Hodge numbers as well as the number of infinitesimal deformations of the vector bundle. We will compare the sum of these numbers on both sides:

$$h^{1,1}_\mathcal{M} + h^{2,1}_\mathcal{M} + h^1_\mathcal{M}(\operatorname{End}(\mathcal{V})) = h^{1,1}_{\widetilde{\mathcal{M}}} + h^{2,1}_{\widetilde{\mathcal{M}}} + h^1_{\widetilde{\mathcal{M}}}(\operatorname{End}(\widetilde{\mathcal{V}})). \tag{9.9}$$

In order to check for these necessary conditions (9.8) and (9.9), for any potentially dual model, we will use the techniques introduced in chapter 7 and furthermore, for simplicity, assume the following statements without testing:

- First of all we assume any dual model $(\widetilde{\mathcal{M}}, \widetilde{\mathcal{V}})$ to have a smooth base as well as a smooth vector bundle. If this is not the case the methods we employ would not give the correct results.

- Secondly we assume stability of the vector bundle $\mathcal{V}$. For generic monad bundles this is in general difficult to check. It implies

$$\begin{aligned} h^0_{\widetilde{\mathcal{M}}}(\widetilde{\mathcal{V}}) &= 0\,, & h^3_{\widetilde{\mathcal{M}}}(\widetilde{\mathcal{V}}) &= 0\,, \\ h^0_{\widetilde{\mathcal{M}}}(\widetilde{\mathcal{V}} \otimes \widetilde{\mathcal{V}}^*) &= 1\,, & h^3_{\widetilde{\mathcal{M}}}(\widetilde{\mathcal{V}} \otimes \widetilde{\mathcal{V}}^*) &= 1\,. \end{aligned} \quad (9.10)$$

- The computation of $h^i_\mathcal{M}(\text{End}(\mathcal{V}))$, explained in 7.4.3, involves a map, for which we assume that it is surjective. This map $\varphi$ appears as the second map in the exact sequence

$$0 \longrightarrow H^0\Big(\mathcal{E}^*_{\widetilde{\mathcal{M}}} \otimes \bigoplus_{l=1}^{n_p} \mathcal{O}_{\widetilde{\mathcal{M}}}(M_l)\Big) \longrightarrow H^0\Big(\bigoplus_{a=1}^{n_\Lambda} \bigoplus_{l=1}^{n_p} \mathcal{O}_{\widetilde{\mathcal{M}}}(M_l - N_a)\Big) \\ \xrightarrow{\varphi} H^0\Big(\bigoplus_{j=1}^{c} \mathcal{O}_{\widetilde{\mathcal{M}}}(M_j)^{\oplus r_{\widetilde{\mathcal{V}}}}\Big) \longrightarrow \cdots \quad (9.11)$$

which arises as an intermediate step in the long exact sequences in cohomology, after writing $\mathcal{V} \otimes \mathcal{V}^*$ via short exact sequences which was derived in its dual version in (7.84). We actually checked for quite a few examples that this holds, but do not have a proof that generically this is the case.

*Remark.* Finally, a comment on the $(0,2)$ moduli space. The number of first order deformations (9.9) is not necessarily equal to the true dimension $\dim(\mathcal{D}(\mathcal{M}, \mathcal{V}))$ of the total moduli space of the theory, as there can be obstructions. Mathematically, this means that there can be complex structure deformations, under which the bundle cannot be kept holomorphic[1]. Physically, this is described by the tree-level four-dimensional superpotential

$$W = \int_\mathcal{M} \Omega_3 \wedge \text{CS}_A \quad (9.12)$$

where $\Omega_3$ denotes the holomorphic $(3,0)$ form on the Calabi-Yau and $\text{CS}_A = \text{tr}(A \wedge dA - \frac{2i}{3} A \wedge A \wedge A)$ the Chern-Simons form of the $SU(n)$ gauge connection $A$. The flat directions of the scalar potential induced by $W$ define the true moduli space of the configuration $(\mathcal{M}, \mathcal{V})$. A non-renormalization theorem states

---

[1] As explained in the physical context for instance in [118], this is captured by the so-called Atiyah-class.

that, beyond this leading order contribution, there can only be non-perturbative corrections from world-sheet instantons. For more information on this important issue, we refer to the literature [119–124].

Unfortunately, the superpotential is hard to compute for a concrete $(0,2)$ model $(\mathcal{M}, \mathcal{V})$. However, we know that at least the independent complex coefficients in the holomorphic sections $G_j$ and $F_a^l$, i.e. the polynomial deformations, keep the vector bundle holomorphic.

## 9.2 Explicit construction of dual $(0,2)$ models

In this section we further generalize the analysis of $(0,2)$ target space dualities presented in [94,95,116,117] and propose a general procedure that can be used to generate dual models from almost any monad over a complete intersection Calabi-Yau base space, not necessarily endowed with a Landau-Ginzburg phase. In particular, one can show that performing this procedure, the anomaly cancellation conditions remain satisfied for the dual models.

### 9.2.1 Outline of the generic construction of dual models

Before we show how to construct dual $(0,2)$ models explicitly, let us outline the generic procedure. As already mentioned, we will start with a smooth $(0,2)$ model $(\mathcal{M}, \mathcal{V})$ for which all anomaly cancellation conditions (2.41) are satisfied. Using the existence of non-geometric phases, where some of the Bosonic fields $p_l$ receive a vev, we perform an exchange of some of the Fermi superfields and the corresponding polynomials. The resulting new GLSM is claimed to be target space dual to the initial one. A necessary condition is that the massless charged matter spectrum and the generic number of massless gauge singlets $D(\mathcal{M}, \mathcal{V})$ should be identical, i.e. (9.8) and (9.9) has to be satisfied (neglecting issues concerning obstructions). More concretely, we follow the procedure:

The procedure:

1. Construct the GLSM phases of a smooth $(0,2)$ model $(\mathcal{M}, \mathcal{V})$.

2. Go to a phase where one of the $p_l$, say $p_1$, is not allowed to vanish and hence obtains a vev $\langle p_1 \rangle$.

3. Perform a rescaling of $k$ Fermi superfields by the constant vev $\langle p_1 \rangle$ and

170                                9. Target Space Dualities in the String Landscape

exchange the role of some $\Lambda^a$ and $\Gamma^j$

$$\widetilde{\Lambda}^{a_i} := \frac{\Gamma^{j_i}}{\langle p_1 \rangle}, \quad \widetilde{\Gamma}^{j_i} := \langle p_1 \rangle \Lambda^{a_i}, \quad \forall i = 1, ..., k,$$

with $\sum_i ||G_{j_i}|| = \sum_i ||F_{a_i}{}^1||$ for anomaly cancellation.

4. Move to a region in the bundle moduli space where the $\Lambda^{a_i}$ only appear in terms with $P_1$ for all $i$. This means that we choose the coefficients in the bundle defining polynomials $F_a{}^l$ such that

$$F_{a_i}{}^l = 0, \quad \forall\, l \neq 1, \; i = 1,...,k.$$

5. Leave the non-geometric phase and define the Fermi superfields of the new GLSM such that each term in the superpotential is $U(1)^r$ gauge invariant. This means

$$||\widetilde{\Lambda}^{a_i}|| = ||\Gamma^{j_i}|| - ||P_1|| \quad \text{and} \quad ||\widetilde{\Gamma}^{j_i}|| = ||\Lambda^{a_i}|| + ||P_1||.$$

6. Returning to a generic point in moduli space defines a new dual $(0, 2)$ GLSM which in a geometric phase corresponds to a different Calabi-Yau/vector bundle configuration $(\widetilde{\mathcal{M}}, \widetilde{\mathcal{V}})$.

### 9.2.2 Explicit procedure

Let us now be more explicit and show how this procedure works in detail. For presentational purpose, we will restrict ourselves to the choice $k = 2$. This is also the case used in performing the $(0, 2)$ landscape analysis to be reported on in section 9.3.

Consider a holomorphic vector bundle $\mathcal{V}$, obtained from a monad over a base Calabi-Yau manifold $\mathcal{M}$ as in (9.5) or (9.3) which we denoted as

$$\mathcal{V}_{N_1,...,N_{n_\Lambda}}[M_1, ..., M_{n_p}] \longrightarrow \mathbb{P}_{Q_1,...,Q_n}[S_1, ..., S_c]. \tag{9.13}$$

It is also assumed that the anomaly cancellation conditions (2.41) are satisfied. We now require that for one specific $M_{l_0}$ there exist two $N_{a_j}$'s such that $N_{a_j}^{(\alpha)} < M_{l_0}^{(\alpha)}$. Let us choose, without loss of generality, $l_0 = 1$ and rearrange the $N's$ in the monad such that they are the first two. Thus the corresponding superpotential

## 9.2. Explicit construction of dual (0,2) models

(2.42) has the form

$$\mathcal{W} = \sum_{j=1}^{c} \Gamma^j G_j + \sum_{a=1}^{2} P_1 \Lambda^a F_a{}^1 + \sum_{a=3}^{n_\Lambda} P_1 \Lambda^a F_a{}^1 + \sum_{l=2}^{n_p} \sum_{a=1}^{n_\Lambda} P_l \Lambda^a F_a{}^l. \qquad (9.14)$$

For the case $n_p = 1$, i.e. a monad $\mathcal{V}_{N_1,...,N_{n_\Lambda}}[M_1]$, the last term would be absent and the GLSM features a Landau-Ginzburg phase in which $p_1$ carries a vacuum expectation value. For the case $n_p = 2$, i.e. $\mathcal{V}_{N_1,...,N_{n_\Lambda}}[M_1, M_2]$, with only a single $U(1)$ gauge symmetry, even though there is no Landau-Ginzburg phase anymore, one may still find a phase in which $p_1$ and $p_2$ cannot vanish simultaneously. This describes a Landau-Ginzburg model fibered over a $\mathbb{P}^1$, parametrized by the homogeneous coordinates $(p_1, p_2)$. Thus, $p_1$ and $p_2$ are not allowed to vanish simultaneously.

Hence, for these two simple cases, one can explicitly identify a phase, in which not all vevs $\langle p_l \rangle$ do vanish. Our dual model generating algorithm starts on a sublocus where a specific vev $\langle p_1 \rangle \neq 0$. This is all we need to perform the desired change of variables. However, since this is a tedious analysis, for the automated landscape study in section 9.3, we did not check for the existence of such a phase for each individual case, but proceeded under the assumption that it exists.

Considering (9.14) and comparing the first sum with the second one, one realizes that they only differ by the additional chiral superfield $P_1$. If one now goes into the aforementioned phase, where $p_1$ obtains a vev, the effective superpotential becomes

$$\mathcal{W} = \sum_{j=1}^{c} \Gamma^j G_j + \sum_{a=1}^{2} \langle p_1 \rangle \Lambda^a F_a{}^1 + \sum_{a=3}^{n_\Lambda} \langle p_1 \rangle \Lambda^a F_a{}^1 + \sum_{l=2}^{n_p} \sum_{a=1}^{n_\Lambda} P_l \Lambda^a F_a{}^l. \qquad (9.15)$$

Now, we want to perform an exchange of two of the Fermi superfields appearing in the first and the second term (9.15). Without loss of generality we choose these two pairs to be $\Gamma^1$, $\Gamma^2$ and $\Lambda^1$, $\Lambda^2$. For this purpose, we first need to move to a region in the bundle moduli space, where the sections $F_a{}^l$ satisfy

$$F_1{}^l = F_2{}^l = 0 \qquad \forall\, l \neq 1. \qquad (9.16)$$

This guarantees that the superpotential takes the restricted form

$$\mathcal{W} = \sum_{j=1}^{c} \Gamma^j G_j + \langle p_1 \rangle \Lambda^1 F_1{}^1 + \langle p_1 \rangle \Lambda^2 F_2{}^1 + \begin{array}{l} \text{terms indep} \\ \text{of } \Gamma^j, \Lambda^1, \Lambda^2 \end{array}. \qquad (9.17)$$

Now the superfields $\Gamma^j$ and $\Lambda^1$, $\Lambda^2$ appear on an equal footing and hence do the homogeneous functions $G_j$, $F_1{}^1$ and $F_2{}^1$. Thus, in this non-geometric phase, their distinctive geometric origin as hypersurface constraints $G_j$ and sections $F_1{}^1$ defining the bundle is completely lost.

Not every such exchange of $\Gamma^1$, $\Gamma^2$ and $\Lambda^1$, $\Lambda^2$ leads to a fully fledged new $(0,2)$ GLSM, after moving away from this special locus in moduli space. For a GLSM the anomaly cancellation conditions (2.41) have to be satisfied. In the following we will describe two different scenarios. The first one corresponds to a consistent exchange of $F$'s and $G$'s where

$$||G_1|| + ||p_1|| \neq 0 \quad \text{and} \quad ||G_2|| + ||p_1|| \neq 0\,, \qquad (9.18)$$

while in the second scenario we will have the situation where

$$||G_1|| + ||p_1|| = 0 \quad \text{and} \quad ||G_2|| + ||p_1|| \neq 0\,. \qquad (9.19)$$

The latter naively leads to Fermi superfields of vanishing charge. We will see that this is not really the case, but that instead for the dual model the number of $U(1)$ gauge symmetries gets increased. Thus, in the geometric phase the dimension of the Kähler moduli space increases. We will find that the Fermi superfield is actually charged under this additional $U(1)$ gauge group.

### Dual models with equal number of $U(1)$ actions.

If we want to consistently exchange $F$'s and $G$'s, we have to make sure that the linear anomaly cancellation condition remains satisfied. For the exchange of two of them, say

$$F_1{}^1,\ F_2{}^1 \quad \rightsquigarrow \quad G_1,\ G_2\,, \qquad (9.20)$$

this requires the following relation of their homogeneous multi-degrees:

$$||F_1{}^1|| + ||F_2{}^1|| = ||G_1|| + ||G_2|| \quad \Rightarrow \quad 2M_1 - N_1 - N_2 = S_1 + S_2\,. \qquad (9.21)$$

As long as $||G_1||$, $||G_2||$ both are not equal to $M_1$, we can perform this exchange without any problem. If there is a phase where $p_1$ is not allowed to vanish, we can write the effective superpotential at low energies by integrating out $p_1$ and moving to the corresponding region in moduli space as seen in (9.17). To make the exchange of the homogeneous polynomials manifest, we have to absorb this

## 9.2. Explicit construction of dual (0,2) models

vev by a rescaling of some of the fields. We obtain the new configuration as

$$\mathcal{W} = \widetilde{\Gamma}^1 \widetilde{G}_1 + \widetilde{\Gamma}^2 \widetilde{G}_2 + \sum_{j=3}^{c} \Gamma^j G_j + \\ \langle p_1 \rangle \widetilde{\Lambda}^1 \widetilde{F}_1{}^1 + \langle p_1 \rangle \widetilde{\Lambda}^2 \widetilde{F}_2{}^1 + \sum_{a=3}^{n_\Lambda} \langle p_1 \rangle \Lambda^a F_a{}^1 + \sum_{l=2}^{n_p} \sum_{a=3}^{n_\Lambda} P_l \Lambda^a F_a{}^l , \qquad (9.22)$$

where we performed rescalings

$$\begin{aligned} \widetilde{\Gamma}^1 &:= \langle p_1 \rangle \Lambda^1 , & \widetilde{\Gamma}^2 &:= \langle p_1 \rangle \Lambda^2 , & \widetilde{\Lambda}^1 &:= \frac{\Gamma^1}{\langle p_1 \rangle}, & \widetilde{\Lambda}^2 &:= \frac{\Gamma^2}{\langle p_1 \rangle}, \\ \widetilde{G}_1 &:= F_1{}^1, & \widetilde{G}_2 &:= F_2{}^1, & \widetilde{F}_1{}^1 &:= G_1, & \widetilde{F}_2{}^1 &:= G_2. \end{aligned} \qquad (9.23)$$

This superpotential (9.22) is identical to the initial one, but arises from a completely different GLSM. At this point we can see that it was essential to move to a specific region of the moduli space, as the rescaling (9.23) would not have been consistent, if there were terms like $\Lambda^1 F_1{}^2$. Since the homogeneous polynomial $F_1{}^2$ might not have the same multi-degree as $F_1{}^1$, the rescaling (9.23) would give rise to a term in the superpotential which is not gauge invariant. The new charges and degrees of the superfields in the GLSM read

$$\mathcal{V}_{\widetilde{N}_1, \widetilde{N}_2, N_3, \dots, N_{n_\Lambda}}[M_1, M_2, \dots, M_{n_p}] \longrightarrow \mathbb{P}_{Q_1, \dots, Q_n}[\widetilde{S}_1, \widetilde{S}_2, S_3, \dots, S_c], \qquad (9.24)$$

with

$$\widetilde{N}_1 := M_1 - S_1, \quad \widetilde{N}_2 := M_2 - S_2, \quad \widetilde{S}_1 := ||F_1{}^1||, \quad \widetilde{S}_2 := ||F_2{}^1||. \qquad (9.25)$$

We prove in appendix A.1 that this $(0,2)$ GLSM fulfills all anomaly cancellation conditions and hence defines a genuine new model. In particular, for the new model one can consider generic points in the moduli space and perform its own phase analysis, i.e. consider the total complexified Kähler moduli space. This also includes the large volume limits of potential geometric phases. There, it describes now topologically distinct Calabi-Yau manifolds equipped with different vector bundles over them.

We were calculating various examples of this kind and found that the following intriguing relation holds in over 90% of them[2]:

$$h_{\mathcal{M}}^{\bullet}(\mathcal{V}) = h_{\widetilde{\mathcal{M}}}^{\bullet}(\widetilde{\mathcal{V}})$$
$$h_{\mathcal{M}}^{1,1} + h_{\mathcal{M}}^{2,1} + h_{\mathcal{M}}^{1}(\text{End}(\mathcal{V})) = h_{\widetilde{\mathcal{M}}}^{1,1} + h_{\widetilde{\mathcal{M}}}^{2,1} + h_{\widetilde{\mathcal{M}}}^{1}(\text{End}(\widetilde{\mathcal{V}})) \,. \tag{9.26}$$

This means that, at least on a dimensional basis, the chiral spectra as well as the number of massless singlets of the two $(0,2)$ models, $(\mathcal{M}, \mathcal{V})$ and $(\widetilde{\mathcal{M}}, \widetilde{\mathcal{V}})$, agree. From the rescalings (9.23), it is clear that the moduli space of $(\mathcal{M}, \mathcal{V})$ is related to the moduli space of $(\widetilde{\mathcal{M}}, \widetilde{\mathcal{V}})$ by an exchange of complex structure and bundle moduli. The Kähler moduli space was rather untouched.

In the following we will describe a way to construct dual $(0,2)$ models in which the Picard group of the ambient space becomes larger so that also the Kähler moduli spaces are non-trivially involved in the duality.

### Dual models with an additional $U(1)$ action.

Let us start again with the monad of the model (9.3) and pick two specific maps $F_a{}^I$, e.g. $F_1{}^1$ and $F_2{}^1$, belonging to $M_1$. Now choose one of the hypersurfaces defining our base, say $S_1$ such that we can find positive degrees $B^{(\alpha)}$ for all $\alpha$ satisfying the multi-degree equation

$$B = ||F_1{}^1|| + ||F_2{}^1|| - S_1 \,. \tag{9.27}$$

We can now introduce a new coordinate $y_1$ with multi-degree $B$ and also a new hypersurface $G^B$ described by a homogeneous polynomial of multi-degree $B$. This means we simply introduce a new Fermi superfield along with a new chiral superfield that have opposite charges. Doing that at the same time does not cause any changes to our model $(\mathcal{M}, \mathcal{V})$ and we can express it as

$$V_{N_1,\ldots,N_{n_\Lambda}}[M_1,\ldots,M_{n_p}] \longrightarrow \mathbb{P}_{Q_1,\ldots,Q_n,B}[S_1,\ldots,S_c,B] \,. \tag{9.28}$$

In case that $S_1 \neq M_1$ and $B \neq M_1$ we can proceed in the same way as in the paragraph above and redefine fields as described there. Thus, we arrive at the following configuration:

$$V_{\widetilde{N}_1,\widetilde{N}_2,N_3,\ldots,N_{n_\Lambda}}[M_1,M_2,\ldots,M_{n_p}] \longrightarrow \mathbb{P}_{Q_1,\ldots,Q_n,B}[\widetilde{S}_1,S_2,\ldots,S_c,\widetilde{B}] \,, \tag{9.29}$$

---

[2]In going through the various Koszul sequences arising for determining the bundle deformations, we were assuming the surjectivity of the map $\varphi$ in (9.11). Moreover, we were also blindly assuming that the new bundle $\widetilde{\mathcal{V}}$ is $\mu$-stable over the new base manifold $\widetilde{\mathcal{M}}$ (9.10). We expect that a mismatch merely indicates that for this specific example one of these assumptions is violated.

## 9.2. Explicit construction of dual (0,2) models

where

$$\widetilde{N}_1 := M_1 - S_1, \quad \widetilde{N}_2 := M_2 - B, \quad \widetilde{S}_1 := ||F_1{}^1||, \quad \widetilde{B} := ||F_2{}^1||. \qquad (9.30)$$

The only new issue is that, as we have introduced a new coordinate $y_1$ with multi-degree $B$, we might get new singularities in the dual model $(\widetilde{\mathcal{M}}, \widetilde{\mathcal{V}})$. These need to be resolved before performing calculations in the large volume limit. In addition, after the resolution of the base, we also have to resolve the bundle without spoiling the anomaly cancellation conditions. How this can be done has been explained in [115].

Applying this procedure for instance to the tangent bundle, i.e. $\mathcal{V} = T_\mathcal{M}$, we encounter the situation that

$$S_1 = M_1 \quad \text{(or equivalently } B = M_1\text{)}. \qquad (9.31)$$

Now performing the same steps as in the last paragraph, we arrive at the following configuration:

$$V_{\widetilde{N}_1, \widetilde{N}_2, N_3, \ldots, N_{n_\Lambda}}[M_1, M_2, \ldots, M_{n_p}] \longrightarrow \mathbb{P}_{Q_1, \ldots, Q_n, B}[\widetilde{S}_1, S_2, \ldots, S_c, \widetilde{B}], \qquad (9.32)$$

where

$$\widetilde{N}_1 := M_1 - S_1 = \vec{0}, \quad \widetilde{N}_2 := M_2 - B, \quad \widetilde{S}_1 := ||F_1{}^1||, \quad \widetilde{B} := ||F_2{}^1||. \qquad (9.33)$$

Thus, in the new model we find the Fermi superfield $\widetilde{\Lambda}^1 := \frac{\Gamma^1}{\langle p_1 \rangle}$ to be uncharged under all of our $U(1)$ symmetries[3].

To proceed, we introduce an additional $U(1)$ gauge symmetry under which the former uncharged Fermi superfield carries a non-vanishing charge. We do that by a formal blow-up of a $\mathbb{P}^1$ with coordinates $y_1, y_2$ so that the charges of the resulting GLSM read

This configuration is equivalent to the initial one. Just eliminate the coordinate $y_1$ via the constraint $G^B = y_1 = 0$ and use the additional $U(1)$ gauge symmetry and the corresponding D-term constraint to fix $y_2$ to a real constant. Then the geometry reduces to $\mathbb{P}_{Q_1, \ldots, Q_n}[S_1, S_2, \ldots, S_c] \times$ pt. So the new configuration also satisfies the anomaly cancellation conditions.

Applying the exchange of $G$'s and $F$'s as in (9.32) and (9.33), the former un-

---
[3]It was argued in [95] that one can employ one of the $n_F$ additional Fermionic gauge symmetries given in (7.61) in order to gauge it away. This works fine for their example but it cannot be used for our case, where the newly introduced field is them self uncharged under all $U(1)$'s.

| $x_1$ | ... | $x_n$ | $y_1$ | $y_2$ | $\Gamma^1$ | ... | $\Gamma^c$ | $\Gamma^B$ |
|---|---|---|---|---|---|---|---|---|
| 0 | ... | 0 | 1 | 1 | 0 | ... | 0 | $-1$ |
| $Q_1$ | ... | $Q_n$ | $B$ | 0 | $-S_1$ | ... | $-S_c$ | $-B$ |

| $\Lambda^1$ | $\Lambda^1$ | ... | $\Lambda^{n_\Lambda}$ | $p_1$ | $p_2$ | ... | $p_{n_p}$ |
|---|---|---|---|---|---|---|---|
| 0 | 0 | ... | 0 | $-1$ | 0 | ... | 0 |
| $N_1$ | $N_2$ | ... | $N_{n_\Lambda}$ | $-M_1$ | $-M_2$ | ... | $-M_{n_p}$ |

,

.

charged Fermi superfield in the dual configuration now carries a non-zero charge under the new $U(1)$. This is what we wanted to achieve and in particular allows us to systematically generate dual models of the heterotic $(\mathcal{M}, T_\mathcal{M})$ models. The data of the GLSM of the dual configuration are listed in table 9.1, for which it is proven in appendix A.1 that they still satisfy all anomaly cancellation conditions (2.41):

| $x_1$ | ... | $x_d$ | $y_1$ | $y_2$ | $\widetilde{\Gamma}^1$ | $\Gamma^2$ | ... | $\Gamma^c$ | $\widetilde{\Gamma}^B$ |
|---|---|---|---|---|---|---|---|---|---|
| 0 | ... | 0 | 1 | 1 | $-1$ | 0 | ... | 0 | $-1$ |
| $Q_1$ | ... | $Q_d$ | $B$ | 0 | $-(M_1 - N_1)$ | $-S_2$ | ... | $-S_c$ | $-(M_1 - N_2)$ |

| $\widetilde{\Lambda}_1$ | $\widetilde{\Lambda}^2$ | ... | $\Lambda^{n_\Lambda}$ | $p_1$ | $p_2$ | ... | $p_{n_p}$ |
|---|---|---|---|---|---|---|---|
| 1 | 0 | ... | 0 | $-1$ | 0 | ... | 0 |
| 0 | $M_2 - B$ | ... | $N_{n_\Lambda}$ | $-M_1$ | $-M_2$ | ... | $-M_{n_p}$ |

,

.

**Table 9.1.:** Charges and degrees of all fields and homogeneous polynomials in the dual model $(\widetilde{S}, \widetilde{V})$

### 9.2.3 Instructive example

Let us present a comparably simple example. Consider the $(0, 2)$ model

$$V_{1,1,1,1,2,2,2}[3,4,3] \longrightarrow \mathbb{P}_{1,1,1,1,2,2,2}[3,4,3], \tag{9.34}$$

where the vector bundle is some deformation of the tangent bundle of the Calabi-Yau manifold $\mathcal{M}$. Since this configuration is singular, we have to resolve it by introducing a new coordinate. After some reordering of the bundle data, this

## 9.2. Explicit construction of dual (0,2) models

yields the following smooth configuration:

| $x_i$ | $\Gamma^j$ |
|---|---|
| 0 0 0 0 1 1 1 1 | −1 −2 −1 |
| 1 1 1 1 2 2 2 0 | −3 −4 −3 |

, (9.35)

| $\Lambda^a$ | $p_l$ |
|---|---|
| 0 1 1 1 0 0 0 1 | −1 −2 −1 |
| 1 2 2 2 1 1 1 0 | −3 −4 −3 |

. (9.36)

We now compute the number of chiral matter zero modes as well as the number of massless singlets

$$h^\bullet_\mathcal{M}(V) = (0, 68, 2, 0),$$
$$h^{1,1}_\mathcal{M} + h^{2,1}_\mathcal{M} + h^1_\mathcal{M}(\mathrm{End}(\mathcal{V})) = 2 + 68 + 140 = 210.$$

(9.37)

Now we can use the procedure from last section. First we introduce a new field $y_1$ which is not charged under the $U(1)$'s we have so far, and introduce a new hypersurface which is also neutral. We formally get the following set of data

| $x_i$ | $\Gamma^j$ |
|---|---|
| 0 0 0 0 1 1 1 1 0 | −1 −2 −1 −0 |
| 1 1 1 1 2 2 2 0 0 | −3 −4 −3 −0 |

. (9.38)

| $\Lambda^a$ | $p_l$ |
|---|---|
| 0 1 1 1 0 0 0 1 | −1 −2 −1 |
| 1 2 2 2 1 1 1 0 | −3 −4 −3 |

. (9.39)

Notice that the homogeneous functions $F_3{}^2$ and $F_4{}^2$ are both of the same multi-degree,

$$||F_3{}^2|| = ||F_4{}^2|| = -||p_2|| - ||\Lambda^3|| = -||p_2|| - ||\Lambda^4|| = \begin{pmatrix} 2-1 \\ 4-2 \end{pmatrix} = \begin{pmatrix} 1 \\ 2 \end{pmatrix}$$ (9.40)

and hence we can exchange the new hypersurface as well as $G_2$ with these two, satisfying

$$||F_3{}^2|| + ||F_4{}^2|| = \begin{pmatrix} 1 \\ 2 \end{pmatrix} + \begin{pmatrix} 1 \\ 2 \end{pmatrix} = \begin{pmatrix} 2 \\ 4 \end{pmatrix} + \begin{pmatrix} 0 \\ 0 \end{pmatrix} = S_2 + B \ . \tag{9.41}$$

From the last section, we know how to exchange these functions and how to redefine the $\Lambda$'s and $\Gamma$'s in order to obtain a sensible new monad. Namely we perform the rescalings

$$\begin{aligned} \widetilde{\Gamma}^2 &:= \langle p_2 \rangle \Lambda^3, & \widetilde{\Gamma}^B &:= \langle p_2 \rangle \Lambda^4, & \widetilde{\Lambda}^3 &:= \frac{\Gamma^2}{\langle p_1 \rangle}, & \widetilde{\Lambda}^4 &:= \frac{\Gamma^B}{\langle p_2 \rangle}, \\ \widetilde{G}_2 &:= F_3{}^2, & \widetilde{G}_B &:= F_4{}^2, & \widetilde{F}_3{}^2 &:= G_2, & \widetilde{F}_4{}^2 &:= G_B, \end{aligned} \tag{9.42}$$

yielding the effective superpotential

$$\begin{aligned} \mathcal{W} = \widetilde{\Gamma}^2 \widetilde{G}_2 + \widetilde{\Gamma}^B \widetilde{G}_B + \sum_{j=1,3} \Gamma^j G_j + \\ \langle p_2 \rangle \widetilde{\Lambda}^3 \widetilde{F}_3{}^2 + \langle p_2 \rangle \widetilde{\Lambda}^4 \widetilde{F}_4{}^2 + \sum_{l=1,3} \sum_{a \neq 3,4} P_l \Lambda^a F_a{}^l \ . \end{aligned} \tag{9.43}$$

The new charges of the constructed model read

$$||\widetilde{\Gamma}^2|| = \begin{pmatrix} -1 \\ -2 \end{pmatrix}, \quad ||\widetilde{\Gamma}^B|| = \begin{pmatrix} -1 \\ -2 \end{pmatrix}, \quad ||\widetilde{\Lambda}^3|| = \begin{pmatrix} 0 \\ 0 \end{pmatrix}, \quad ||\widetilde{\Lambda}^4|| = \begin{pmatrix} 2 \\ 4 \end{pmatrix}, \tag{9.44}$$

$$||\widetilde{G}_2|| = \begin{pmatrix} 1 \\ 2 \end{pmatrix}, \quad ||\widetilde{G}_B|| = \begin{pmatrix} 1 \\ 2 \end{pmatrix}, \quad ||\widetilde{F}_3{}^2|| = \begin{pmatrix} 2 \\ 4 \end{pmatrix}, \quad ||\widetilde{F}_4{}^2|| = \begin{pmatrix} 0 \\ 0 \end{pmatrix}. \tag{9.45}$$

We realize that one of the new $\Lambda$'s is uncharged under both $U(1)$'s. Thus, we introduce a new $U(1)$ along with a new coordinate in the base, which gives all new fields a charge. Doing that in the way explained above and going back to a

## 9.2. Explicit construction of dual (0,2) models

generic point in moduli space, we arrive at the GLSM

| $x_i$ | | | | | | | | | | $\Gamma^j$ | | | |
|---|---|---|---|---|---|---|---|---|---|---|---|---|---|
| 0 | 0 | 0 | 0 | 0 | 0 | 0 | 1 | 1 | | $-0$ | $-1$ | $-0$ | $-1$ |
| 0 | 0 | 0 | 0 | 1 | 1 | 1 | 1 | 0 | 0 | $-1$ | $-1$ | $-1$ | $-1$ |
| 1 | 1 | 1 | 1 | 2 | 2 | 2 | 0 | 0 | | $-3$ | $-2$ | $-3$ | $-2$ |

(9.46)

| $\Lambda^a$ | | | | | | | | $p_l$ | | |
|---|---|---|---|---|---|---|---|---|---|---|
| 0 | 0 | 1 | 0 | 0 | 0 | 0 | 0 | $-0$ | $-1$ | $-0$ |
| 0 | 1 | 0 | 2 | 0 | 0 | 0 | 1 | $-1$ | $-2$ | $-1$ |
| 1 | 2 | 0 | 4 | 1 | 1 | 1 | 0 | $-3$ | $-4$ | $-3$ |

(9.47)

As was generically shown, this configuration satisfies the conditions (2.41) and we obtain the following topological data:

$$h^{\bullet}_{\widetilde{\mathcal{M}}}(\widetilde{\mathcal{V}}) = (0, 68, 2, 0),$$
$$h^{1,1}_{\widetilde{\mathcal{M}}} + h^{2,1}_{\widetilde{\mathcal{M}}} + h^{1}_{\widetilde{\mathcal{M}}}(\mathrm{End}(\widetilde{\mathcal{V}})) = 3 + 51 + 156 = 210.$$

(9.48)

Comparison to the data (9.37) tells us that the number of chiral zero modes did not change and, even though the individual Hodge numbers changed, the total number of first order deformations stayed the same.

This was just one possible choice of a pair of $F$'s and $G$'s, but actually not the only one. We could for instance exchange a different pair, which involves a redefinition of $\Lambda^1$ and $\Lambda^2$ rather than $\Lambda^3$ and $\Lambda^4$. In this case we finally obtain the GLSM

| $x_i$ | | | | | | | | | | $\Gamma^j$ | | | |
|---|---|---|---|---|---|---|---|---|---|---|---|---|---|
| 0 | 0 | 0 | 0 | 0 | 0 | 0 | 1 | 1 | | $-1$ | $-0$ | $-0$ | $-1$ |
| 0 | 0 | 0 | 0 | 1 | 1 | 1 | 1 | 0 | 0 | $-1$ | $-2$ | $-1$ | $-0$ |
| 1 | 1 | 1 | 1 | 2 | 2 | 2 | 0 | 0 | | $-2$ | $-4$ | $-3$ | $-1$ |

(9.49)

| $\Lambda^a$ | | | | | | | | $p_l$ | | |
|---|---|---|---|---|---|---|---|---|---|---|
| 1 | 0 | 0 | 0 | 0 | 0 | 0 | 0 | $-1$ | $-0$ | $-0$ |
| 0 | 1 | 1 | 1 | 0 | 0 | 0 | 1 | $-1$ | $-2$ | $-1$ |
| 0 | 3 | 2 | 2 | 1 | 1 | 1 | 0 | $-3$ | $-4$ | $-3$ |

(9.50)

and find for the massless spectrum

$$\begin{aligned} h^{\bullet}_{\widehat{\mathcal{M}}}(\widehat{\mathcal{V}}) &= (0, 68, 2, 0)\,, \\ h^{1,1}_{\widehat{\mathcal{M}}} + h^{2,1}_{\widehat{\mathcal{M}}} + h^{1}_{\widehat{\mathcal{M}}}(\text{End}(\widehat{\mathcal{V}})) &= 3 + 63 + 144 = 210\,. \end{aligned} \quad (9.51)$$

### 9.2.4 The dual base via a conifold transition

The methods described in section 9.2.2 are applicable to almost any $(0,2)$ GLSM. Starting with a model that admits a $(2,2)$ locus, namely a heterotic model with standard embedding, we have seen that in the dual $(0,2)$ model one has to introduce a new $\mathbb{P}^1$. This results in a base Calabi-Yau manifold $\widetilde{\mathcal{M}}$ whose Kähler moduli space has a higher dimension than the original one for $\mathcal{M}$.

The question now is, what the geometric relation between the two Calabi-Yau manifolds $\mathcal{M}$ and $\widetilde{\mathcal{M}}$ is. As already observed for a specific example in [116, 117], $\widetilde{\mathcal{M}}$ seems to be connected to $\mathcal{M}$ via a conifold transition. Let us explain this for our more generic situation in more detail.

#### Standard embedding:

Let us first consider a Calabi-Yau manifold $\mathcal{M}$ with a holomorphic vector bundle $\mathcal{V}$ which is a deformation of the tangent bundle $T_\mathcal{M}$. As before, let $G_1, \ldots, G_c$ be the intersecting hypersurfaces and $x_1, \ldots, x_n$ the homogeneous coordinates of the ambient space. Let us pick an arbitrary hypersurface, say $G_1$, and move to a specific region in the complex structure moduli space where we can write this surface as a combination of polynomials of lower degree:

$$G_1 = x_1 \, F_1{}^1 - x_2 \, F_2{}^1 = 0\,, \quad (9.52)$$

where $||F_i^j|| = ||G_j|| - ||x_i||$. At this point in complex structure moduli space the manifold develops a conifold singularity. It is well known that it can be resolved via a small resolution [125]. This is described by introducing two new coordinates, $y_1$ and $y_2$, parameterizing a $\mathbb{P}^1$ satisfying the two hypersurface constraints

$$\begin{aligned} \widetilde{G}_1 &:= y_1 \, x_1 + y_2 \, F_2{}^1 = 0\,, \\ \widetilde{G}_B &:= y_1 \, x_2 + y_2 \, F_1{}^1 = 0 \end{aligned} \quad (9.53)$$

## 9.2. Explicit construction of dual (0,2) models

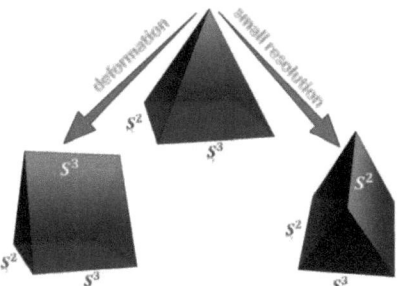

**Figure 9.1.:** Pictorial illustration of the conifold transition via a deformation on the one side and a small resolution on the other side. This corresponds basically to the scenario, where the CY hypersurface is given by the quintic polynomial $G = (\sum_{i=1}^{5} x_i^5) - 5 \cdot c \cdot x_1 x_2 x_3 x_4 x_5$ in $\mathbb{P}^4$. If we deform this polynomial such that $c = 1$, this submanifold becomes singular and can locally be described as a cone over the base $S^2 \times S^3$ [125].

which can be written as[4]

$$M \cdot \begin{pmatrix} y_1 \\ y_2 \end{pmatrix} = 0, \qquad M := \begin{pmatrix} x_1 & F_2{}^1 \\ x_2 & F_1{}^1 \end{pmatrix}. \tag{9.54}$$

Since $y_1$ and $y_2$ are not allowed to vanish simultaneously, the conifold (9.52) is recovered from $\det(M) = 0$. This is the locus to which the resolved space degenerates in the limit of vanishing size of the $\mathbb{P}^1$. The procedure of deformation and small resolution can be seen pictorially in figure 9.1. For the degrees of the two new hypersurfaces one obtains

$$||\widetilde{G}_1|| = \begin{pmatrix} 1 \\ ||G_1|| - ||x_2|| \end{pmatrix} = \begin{pmatrix} 1 \\ ||F_2{}^1|| \end{pmatrix}, \tag{9.55}$$

$$||\widetilde{G}_B|| = \begin{pmatrix} 1 \\ ||G_1|| - ||x_1|| \end{pmatrix} = \begin{pmatrix} 1 \\ ||F_1{}^1|| \end{pmatrix}. \tag{9.56}$$

For the $(0,2)$ model, where $\mathcal{V}$ is a deformation of the tangent bundle $T_\mathcal{M}$, the degree of the $F_1{}^1, F_2{}^1$ in (9.52) is equal to the degree of the $F_a{}^l$ in the monad (7.61). Therefore the conifold transition is equivalent to the transformation of

---

[4]In the literature, often the resolution $M \to M^T$ is considered.

the base described in 9.2.2. Thus, for the target space dual pair

$$(\mathcal{M}, \mathcal{V}) \rightsquigarrow (\widetilde{\mathcal{M}}, \widetilde{\mathcal{V}}), \qquad (9.57)$$

where $\mathcal{V}$ a deformation of $T_\mathcal{M}$, the two base manifolds $\mathcal{M}$ and $\widetilde{\mathcal{M}}$ are connected via a conifold transition. In contrast to the conifold transition for type II superstrings, i.e. for $(2,2)$ models, here $\widetilde{\mathcal{V}} \neq T_{\widetilde{\mathcal{M}}}$. Moreover, we are not claiming that there is a physically smooth transition between the two configurations. On the contrary, our point is that the two $(0,2)$ models are isomorphic descriptions of the same stringy geometry.

As an example from [125], consider the Calabi-Yau manifold given by the complete intersection of two surfaces of degree 4 and 2 in $\mathbb{P}^5$, i.e. $\mathcal{M} = \mathbb{P}^5[4,2]$. Here we have

$$\begin{aligned} h^\bullet_\mathcal{M}(T_\mathcal{M}) &= (0, 89, 1, 0) \\ h^{1,1}_\mathcal{M} + h^{2,1}_\mathcal{M} + h^1_\mathcal{M}(\mathrm{End}(T_\mathcal{M})) &= 1 + 89 + 190 = 280 \,. \end{aligned} \qquad (9.58)$$

It has been shown in [125] that via a conifold transition, this manifold is connected to the new Calabi-Yau $\widetilde{\mathcal{M}}$ defined as

| $x_i$ | | | | | | | | $\Gamma^j$ | | |
|---|---|---|---|---|---|---|---|---|---|---|
| 0 | 0 | 0 | 0 | 0 | 0 | 1 | 1 | $-0$ | $-1$ | $-1$ |
| 1 | 1 | 1 | 1 | 1 | 1 | 0 | 0 | $-4$ | $-1$ | $-1$ |

which has Hodge numbers $(h^{2,1}_{\widetilde{\mathcal{M}}}, h^{1,1}_{\widetilde{\mathcal{M}}}) = (86, 2)$ and $h^1_{\widetilde{\mathcal{M}}}(\mathrm{End}(T_{\widetilde{\mathcal{M}}})) = 188$ bundle moduli which gives a total number of 276 massless singlets for the corresponding $(2,2)$ model. In our situation, the dual vector bundle is different and given by

| $\Lambda^a$ | | | | | | $p_l$ | |
|---|---|---|---|---|---|---|---|
| 0 | 0 | 0 | 0 | 1 | 0 | $-0$ | $-1$ |
| 1 | 1 | 1 | 1 | 0 | 2 | $-4$ | $-2$ |

We compute

$$\begin{aligned} h^\bullet_{\widetilde{\mathcal{M}}}(\widetilde{\mathcal{V}}) &= (0, 89, 1, 0) \,, \\ h^{1,1}_{\widetilde{\mathcal{M}}} + h^{2,1}_{\widetilde{\mathcal{M}}} + h^1_{\widetilde{\mathcal{M}}}(\mathrm{End}(\widetilde{\mathcal{V}})) &= 2 + 86 + 192 = 280 \end{aligned} \qquad (9.59)$$

so that the number of complex structure moduli decreased by three, whereas the number of bundle and Kähler moduli increased by two and one, respectively.

## 9.2. Explicit construction of dual (0,2) models

### Generic case:

Turning now to the case of a non-standard embedding, the story changes only slightly. If we can find a point in the complex structure and bundle moduli space such that two of the $F$'s actually appear in one and the same $G$ as[5]

$$G_1 = U_1 F_1^1 - U_2 F_2^1 = 0, \qquad (9.60)$$

where the $U_i$ are homogeneous polynomials such that

$$||U_i|| = S_1 - M_1 + N_i, \qquad \text{for } i = 1, 2. \qquad (9.61)$$

As before, (9.60) defines a conifold singularity, which can be resolved by blowing up $\mathbb{P}^1$s over the nodal points. This is described by introducing two new coordinates $y_1, y_2$ parameterizing the $\mathbb{P}^1$ and the two hypersurfaces

$$\begin{aligned}\widetilde{G}_1 &:= y_1 U_1 + y_2 F_2^1 = 0, \\ \widetilde{G}_B &:= y_1 U_2 + y_2 F_1^1 = 0\end{aligned} \qquad (9.62)$$

which can also be written as

$$M \cdot \begin{pmatrix} y_1 \\ y_2 \end{pmatrix} = 0, \qquad M := \begin{pmatrix} U_1 & F_2^1 \\ U_2 & F_1^1 \end{pmatrix}. \qquad (9.63)$$

The new degrees of the new coordinates and constraints are given as

$$\begin{aligned}||y_2|| &= \begin{pmatrix} 1 \\ \vec{0} \end{pmatrix}, & ||y_1|| &= \begin{pmatrix} 1 \\ ||F_1^1|| + ||F_2^1|| - S_1 \end{pmatrix}, \\ ||\widetilde{G}_1|| &= \begin{pmatrix} 1 \\ ||F_2^1|| \end{pmatrix}, & ||\widetilde{G}_B|| &= \begin{pmatrix} 1 \\ ||F_1^1|| \end{pmatrix}.\end{aligned} \qquad (9.64)$$

This is precisely what we obtained for the dual Calabi-Yau manifold $\widetilde{\mathcal{M}}$ in subsection 9.2.2. Therefore, also for this more generic case the two base manifolds are connected by a conifold transition.

---

[5]This will always happen as long as the degree of the $F$'s are less or equal to the degree of the $G$.

## 9.2.5 Chains of dual models

As we have explained in the beginning of this section, the proposed construction of potentially dual models is pretty independent of the choice of the monad, unless the data is chosen so badly, that no exchange satisfying (9.21) is possible. Therefore, one is free to iterate the procedure to produce dual $(0,2)$ models, until one arrives at a monad already obtained before. Depending on the initial GLSM data of $(\mathcal{M}_0, \mathcal{V}_0)$, this can lead to quite a number of dual configurations $(\mathcal{M}_i, \mathcal{V}_i)$, $i = 1, \ldots, m$. In all cases investigated the number $m$ is finite.

To show one example we choose a product of projective spaces, where the hypersurfaces have multi-degrees containing only 1's or 0's. The starting point is given by the $(2,2)$ model

| $x_i$ | $-\Gamma^j$ | $\Lambda^a$ | $-p_l$ |
|---|---|---|---|
| $\mathbb{P}^2$ | 0 0 1 1 1 | 0 0 0 0 0 1 0 1 0 1 | 0 0 1 1 1 |
| $\mathbb{P}^2$ | 1 1 1 0 0 | 0 1 0 1 0 1 0 0 0 0 0 | 1 1 1 0 0 |
| $\mathbb{P}^4$ | 1 1 1 1 1 | 1 0 1 0 1 0 0 1 0 1 0 | 1 1 1 1 1 |

,

where the first column is meant to be $\mathbb{P}^2 \times \mathbb{P}^2 \times \mathbb{P}^4$. The topological data of this configuration is given by

$$h^\bullet_\mathcal{M}(\mathcal{V}) = (0, 44, 3, 0),$$
$$h^{1,1}_\mathcal{M} + h^{2,1}_\mathcal{M} + h^1_\mathcal{M}(\text{End}(\mathcal{V})) = 3 + 44 + 48 = 95. \quad (9.65)$$

For this example we can apply the procedure five times until we do not obtain anything new anymore. All these new monads are topologically different and all have the same chiral spectrum as the initial one:

| Nr. | $h^1_\mathcal{M}(\mathcal{V})$ | $h^2_\mathcal{M}(\mathcal{V})$ | $h^{1,1}_\mathcal{M}$ | $h^{2,1}_\mathcal{M}$ | $h^1_\mathcal{M}(\text{End}(\mathcal{V}))$ | $D(\mathcal{M}, \mathcal{V})$ |
|---|---|---|---|---|---|---|
| 1 | 44 | 3 | 4 | 42 | 49 | 95 |
| 2 | 44 | 3 | 5 | 40 | 50 | 95 |
| 3 | 44 | 3 | 6 | 38 | 51 | 95 |
| 4 | 44 | 3 | 7 | 36 | 52 | 95 |

The defining data can be derived to be

## 9.2. Explicit construction of dual (0,2) models

Model 0:

| | | | |
|---|---|---|---|
| $\mathbb{P}^2$ | 0 0 1 1 1 | 0 0 0 0 0 0 0 1 1 1 | 0 0 1 1 1 |
| $\mathbb{P}^2$ | 1 1 1 0 0 | 0 0 0 0 0 1 1 1 0 0 0 | 1 1 1 0 0 |
| $\mathbb{P}^4$ | 1 1 1 1 1 | 1 1 1 1 1 0 0 0 0 0 0 | 1 1 1 1 1 |

Model 1:

| | | | |
|---|---|---|---|
| $\mathbb{P}^1$ | 1 0 0 0 0 1 | 1 0 0 0 0 0 0 0 0 0 0 | 1 0 0 0 0 |
| $\mathbb{P}^2$ | 0 0 1 1 1 0 | 0 0 0 0 0 0 0 0 1 1 1 | 0 0 1 1 1 |
| $\mathbb{P}^2$ | 1 1 1 0 0 0 | 0 0 0 0 0 1 1 1 0 0 0 | 1 1 1 0 0 |
| $\mathbb{P}^4$ | 0 1 1 1 1 1 | 0 1 1 1 1 1 0 0 0 0 0 | 1 1 1 1 1 |

Model 2:

| | | | |
|---|---|---|---|
| $\mathbb{P}^1$ | 0 1 0 0 0 0 1 | 0 1 0 0 0 0 0 0 0 0 0 | 0 1 0 0 0 |
| $\mathbb{P}^1$ | 1 0 0 0 0 1 0 | 1 0 0 0 0 0 0 0 0 0 0 | 1 0 0 0 0 |
| $\mathbb{P}^2$ | 0 0 1 1 1 0 0 | 0 0 0 0 0 0 0 0 1 1 1 | 0 0 1 1 1 |
| $\mathbb{P}^2$ | 1 1 1 0 0 0 0 | 0 0 0 0 0 1 1 1 0 0 0 | 1 1 1 0 0 |
| $\mathbb{P}^4$ | 0 0 1 1 1 1 1 | 0 0 1 1 1 1 1 0 0 0 0 | 1 1 1 1 1 |

Model 3:

| | | | |
|---|---|---|---|
| $\mathbb{P}^1$ | 0 0 0 1 0 0 0 1 | 0 0 0 1 0 0 0 0 0 0 0 | 0 0 0 1 0 |
| $\mathbb{P}^1$ | 0 1 0 0 0 0 1 0 | 0 1 0 0 0 0 0 0 0 0 0 | 0 1 0 0 0 |
| $\mathbb{P}^1$ | 1 0 0 0 0 1 0 0 | 1 0 0 0 0 0 0 0 0 0 0 | 1 0 0 0 0 |
| $\mathbb{P}^2$ | 0 0 1 1 1 0 0 0 | 0 0 0 0 0 0 0 0 1 1 1 | 0 0 1 1 1 |
| $\mathbb{P}^2$ | 1 1 1 0 0 0 0 0 | 0 0 0 0 0 1 1 1 0 0 0 | 1 1 1 0 0 |
| $\mathbb{P}^4$ | 0 0 1 0 1 1 1 1 | 0 0 1 0 1 1 1 0 0 1 0 | 1 1 1 1 1 |

Model 4:

| | | | |
|---|---|---|---|
| $\mathbb{P}^1$ | 0 0 0 0 1 0 0 0 1 | 0 0 0 0 1 0 0 0 0 0 | 0 0 0 0 1 |
| $\mathbb{P}^1$ | 0 0 0 1 0 0 0 1 0 | 0 0 0 1 0 0 0 0 0 0 | 0 0 0 1 0 |
| $\mathbb{P}^1$ | 0 1 0 0 0 0 1 0 0 | 0 1 0 0 0 0 0 0 0 0 | 0 1 0 0 0 |
| $\mathbb{P}^1$ | 1 0 0 0 0 1 0 0 0 | 1 0 0 0 0 0 0 0 0 0 | 1 0 0 0 0 |
| $\mathbb{P}^2$ | 0 0 1 1 1 0 0 0 0 | 0 0 0 0 0 0 0 1 1 1 | 0 0 1 1 1 |
| $\mathbb{P}^2$ | 1 1 1 0 0 0 0 0 0 | 0 0 0 0 0 1 1 1 0 0 | 1 1 1 0 0 |
| $\mathbb{P}^4$ | 0 0 1 0 0 1 1 1 1 | 0 0 1 0 0 1 1 0 0 1 1 | 1 1 1 1 1 |

## 9.2.6 Honest $(0,2)$ models with structure group $SU(n)$

In subsections 9.2.3 and 9.2.5 we gave different examples for target space dual configurations. Nevertheless we only considered examples where the starting point was an $SU(3)$ model derived as deformation of the tangent bundle. We now want to consider more generic examples of the duality, namely "honest" $(0,2)$ models that are not given by the deformation of the tangent bundle. Each type of structure group $SU(3)$, $SU(4)$ and $SU(5)$ will be covered separately. In contrast to what we showed before, here we want to give different examples that do arise from an exact monad introduced in (7.43)

$$0 \to \mathcal{V} \xrightarrow{\iota} \bigoplus_{a=1}^{n_\Lambda} \mathcal{O}_{\mathcal{M}}(N_a) \xrightarrow{\otimes F_a{}^l} \bigoplus_{l=1}^{n_p} \mathcal{O}_{\mathcal{M}}(M_l) \to 0 \,,$$

and hence are given by the kernel of the map $F_a{}^l$

$$\mathcal{V} = \ker\left(F_a{}^l\right) . \tag{9.66}$$

The way to generate the dual models of such a monad remains the same.

**Example 9.2.1** (Example for an $SU(3)$-model)**.** We start with an $SU(3)$ example which consists of a holomorphic vector bundle over a codimension two complete intersection Calabi-Yau space. In this example here we are not dealing with a $(0,2)$ model which is a deformation of the tangent bundle but a completely independent monad. Furthermore we will see that the base will not transform via a conifold transition. Rather in the beginning the ambient variety will remain untouched and only a different set of hypersurfaces will be chosen, also resulting in a topology change of the base. Finally through the exchange of those specific hypersurfaces we will see that in fact the ambient space topology will be changed

## 9.2. Explicit construction of dual (0,2) models

after all. The model data is given by

| $x_i$ |   |   |   |   |   |   ‖ $\Gamma^j$ |    |
|---|---|---|---|---|---|---|---|---|
| 0 | 0 | 0 | 1 | 1 | 1 | 1 ‖ $-2$ | $-2$ |
| 1 | 1 | 1 | 2 | 2 | 2 | 0 ‖ $-4$ | $-5$ |

| $\Lambda^a$ |   |   |   ‖ $p_l$ |
|---|---|---|---|---|
| 1 | 0 | 0 | 2 ‖ $-3$ |
| 0 | 1 | 1 | 6 ‖ $-8$ |

(9.67)

Employing our implementation cohomCalg Koszul extension [69] for this matter we find

$$h^{\bullet}_{\mathcal{M}}(\mathcal{V}) = (0, 120, 0, 0),$$
$$h^{1,1}_{\mathcal{M}} + h^{2,1}_{\mathcal{M}} + h^1_{\mathcal{M}}(\operatorname{End}(\mathcal{V})) = 2 + 68 + 322 = 392.$$

(9.68)

In order to see our freedom of consistently exchanging hypersurface equations with bundle maps in the monad as described in the last section we explicitly write down the multi-degrees of the corresponding generic homogeneous functions. Using that

$$||F_a{}^l|| = -||p_l|| - ||\Lambda^a||,$$

(9.69)

for the only choice $l = 1$ they read

$$||G_1|| = \begin{pmatrix} 2 \\ 4 \end{pmatrix}, ||G_2|| = \begin{pmatrix} 2 \\ 5 \end{pmatrix},$$

(9.70)

$$||F_1{}^1|| = \begin{pmatrix} 2 \\ 8 \end{pmatrix}, ||F_2{}^1|| = \begin{pmatrix} 3 \\ 7 \end{pmatrix}, ||F_3{}^1|| = \begin{pmatrix} 3 \\ 7 \end{pmatrix}, ||F_4{}^1|| = \begin{pmatrix} 1 \\ 2 \end{pmatrix}.$$

(9.71)

Here we can already see that the sum of the degrees of the two hypersurfaces equals the sum of the degree of the third and the fourth $F$. From the last section, we know how to exchange these functions and how to redefine the $\Lambda$'s and $\Gamma$'s in order to obtain a sensible new monad. Namely we perform the rescalings

$$\widetilde{\Gamma}^1 := \langle p_1 \rangle \Lambda^3, \quad \widetilde{\Gamma}^B := \langle p_1 \rangle \Lambda^4, \quad \widetilde{\Lambda}^3 := \frac{\Gamma^1}{\langle p_1 \rangle}, \quad \widetilde{\Lambda}^4 := \frac{\Gamma^B}{\langle p_1 \rangle},$$
$$\widetilde{G}_1 := F_3{}^1, \quad \widetilde{G}_2 := F_4{}^1, \quad \widetilde{F}_3{}^1 := G_1, \quad \widetilde{F}_4{}^1 := G_2,$$

(9.72)

yielding the effective superpotential

$$\mathcal{W} = \widetilde{\Gamma}^1 \widetilde{G}_1 + \widetilde{\Gamma}^2 \widetilde{G}_2 + \langle p_1 \rangle \left( \widetilde{\Lambda}^3 \widetilde{F}_3{}^1 + \widetilde{\Lambda}^4 \widetilde{F}_4{}^1 + \Lambda^1 F_1{}^1 + \Lambda^2 F_2{}^1 \right).$$

(9.73)

The new charges of the constructed model read

$$||\widetilde{\Gamma}^1|| = \begin{pmatrix} -3 \\ -7 \end{pmatrix}, \quad ||\widetilde{\Gamma}^2|| = \begin{pmatrix} -1 \\ -2 \end{pmatrix}, \quad ||\widetilde{\Lambda}^3|| = \begin{pmatrix} 1 \\ 4 \end{pmatrix}, \quad ||\widetilde{\Lambda}^4|| = \begin{pmatrix} 1 \\ 3 \end{pmatrix},$$
$$||\widetilde{G}_1|| = \begin{pmatrix} 3 \\ 7 \end{pmatrix}, \quad ||\widetilde{G}_2|| = \begin{pmatrix} 1 \\ 2 \end{pmatrix}, \quad ||\widetilde{F}_3{}^1|| = \begin{pmatrix} 2 \\ 4 \end{pmatrix}, \quad ||\widetilde{F}_4{}^1|| = \begin{pmatrix} 2 \\ 5 \end{pmatrix}$$
(9.74)

and hence going back to the geometric phase we obtain the new base with a new vector bundle. We notice that the new configuration can be rewritten in a slightly simpler way. The new hypersurface $\widetilde{G}_2$ has precisely the same degree as the divisor $\{x_4 = 0\}$ and therefore the corresponding constraining equation to the ambient space simply removes this coordinate from the configuration and we obtain

| $x_i$ | | | | | | $\Gamma^j$ | | $\Lambda^a$ | | | | $p_l$ |
|---|---|---|---|---|---|---|---|---|---|---|---|---|
| 0 | 0 | 0 | 1 | 1 | 1 | −3 | | 1 | 0 | 1 | 1 | −3 |
| 1 | 1 | 1 | 2 | 2 | 0 | −7 | | 0 | 1 | 4 | 3 | −8 |

(9.75)

As was generically shown, this configuration still satisfies the conditions (2.41) and we obtain the following topological data:

$$h^\bullet_{\widetilde{\mathcal{M}}}(\widetilde{\mathcal{V}}) = (0, 120, 0, 0),$$
$$h^{1,1}_{\widetilde{\mathcal{M}}} + h^{2,1}_{\widetilde{\mathcal{M}}} + h^1_{\widetilde{\mathcal{M}}}(\mathrm{End}(\widetilde{\mathcal{V}})) = 2 + 95 + 295 = 392.$$
(9.76)

If we compare this with the result we obtained in (9.68), we see that the number of chiral zero modes did not change and the total number of first order deformations stayed the same even though the Hodge number $h^{2,1}$ changed drastically.

Let us once more put some emphasis on the fact that we started up with a base manifold that was of codimension two and due to the exchange ended up with a simpler space given by a codimension one Calabi-Yau manifold. Similarly, as we will see in the next example this can also happen the other way round resulting in an increase of the codimension. Also the number of $C^*$ actions can change which will be shown in the following examples, too.

**Example 9.2.2** (An example for an $SU(4)$-model). Next we present an example of a dual pair of heterotic $(0, 2)$ models that give rise to gauge group $SO(10)$ in four dimensions and hence are equipped with a rank 4 vector bundle. The model is again not a deformation of the tangent bundle. The base is the complete intersection of a generic quartic and homogeneous degree hypersurface two inside

## 9.2. Explicit construction of dual (0,2) models

$\mathbb{P}^5$. The defining data can be read off in the following table:

| $x_i$ | $\Gamma^j$ | | $\Lambda^a$ | | | | | | | $p_l$ | | |
|---|---|---|---|---|---|---|---|---|---|---|---|---|
| $\mathbb{P}^5$ | $-2$ | $-4$ | 1 | 1 | 1 | 1 | 1 | 1 | 1 | $-3$ | $-2$ | $-2$ |

(9.77)

Clearly this model is anomaly free, i.e. it satisfies (2.41) and one can also show that the vector bundle is also stable[6]. It is sometimes also referred to as a positive monad, since all line bundles involved have positive degree. It has the following topological data:

$$h^{\bullet}_{\widetilde{\mathcal{M}}}(\widetilde{\mathcal{V}}) = (0, 48, 0, 0),$$
$$h^{1,1}_{\widetilde{\mathcal{M}}} + h^{2,1}_{\widetilde{\mathcal{M}}} + h^{1}_{\widetilde{\mathcal{M}}}(\text{End}(\widetilde{\mathcal{V}})) = 1 + 89 + 159 = 249.$$

(9.78)

Before we move on, we introduce a new coordinate along with a new hypersurface to the model. Doing that at the same time does not change the model at all. In order to perform the exchange of polynomials $F$ and $G$, we have to go to a certain region of the moduli space, exchange them and go back to the generic region in the dual configuration. The resulting base manifold can then be obtained as the conifold transition of the initial base space. The full model is given by

| $x_i$ | $\Gamma^j$ | | | $\Lambda^a$ | | | | | | | $p_l$ | | |
|---|---|---|---|---|---|---|---|---|---|---|---|---|---|
| $\mathbb{P}^1$ | $-1$ | 0 | $-1$ | 0 | 0 | 0 | 1 | 0 | 0 | 0 | 0 | $-1$ | 0 |
| $\mathbb{P}^5$ | $-1$ | $-4$ | $-1$ | 1 | 1 | 1 | 0 | 2 | 1 | 1 | $-3$ | $-2$ | $-2$ |

(9.79)

and its topology satisfies the necessary duality check of coinciding spectrum and moduli space dimensions:

$$h^{\bullet}_{\widetilde{\mathcal{M}}}(\widetilde{\mathcal{V}}) = (0, 48, 0, 0),$$
$$h^{1,1}_{\widetilde{\mathcal{M}}} + h^{2,1}_{\widetilde{\mathcal{M}}} + h^{1}_{\widetilde{\mathcal{M}}}(\text{End}(\widetilde{\mathcal{V}})) = 2 + 86 + 161 = 249.$$

(9.80)

**Example 9.2.3** (An example for an $SU(5)$-model). Finally let us quickly state a different bundle over the same base from example 9.2.2. We modify it such that

---

[6]By Hoppe's criterion for the initial bundle in this and the next example, to prove stability it suffices to show that $h^0(\mathcal{M}; \Lambda^k \mathcal{V}) = 0 \ \forall k < \text{rk}\mathcal{V}$ and is relatively straight forward in these cases.

it has no longer $SU(4)$ but rather $SU(5)$ structure. It is given by

| $x_i$ | $\Gamma^j$ | | $\Lambda^a$ | | | | | | | | $p_l$ | | |
|---|---|---|---|---|---|---|---|---|---|---|---|---|---|
| $\mathbb{P}^5$ | $-2$ | $-4$ | 1 | 1 | 1 | 1 | 1 | 1 | 1 | 1 | $-3$ | $-3$ | $-2$ |

. (9.81)

Since we have still three chiral fields $p_l$ but eight Fermi fields $\Lambda^a$, we end up with a rank five vector bundle and hence yielding an $SU(5)$ gauge group in the four-dimensional theory. The spectrum and the dimension of the moduli space for this model can be calculated as

$$h^\bullet_{\widetilde{\mathcal{M}}}(\widetilde{\mathcal{V}}) = (0, 72, 0, 0)\,,$$
$$h^{1,1}_{\widetilde{\mathcal{M}}} + h^{2,1}_{\widetilde{\mathcal{M}}} + h^1_{\widetilde{\mathcal{M}}}(\mathrm{End}(\widetilde{\mathcal{V}})) = 1 + 89 + 288 = 378\,. \quad (9.82)$$

The dual base is again given by the same conifold transition as in the last paragraph. Altogether we get

| $x_i$ | $\Gamma^j$ | | | $\Lambda^a$ | | | | | | | | $p_l$ | | |
|---|---|---|---|---|---|---|---|---|---|---|---|---|---|---|
| $\mathbb{P}^1$ | $-1$ | $0$ | $-1$ | 0 | 0 | 0 | 0 | 0 | 0 | 1 | 0 | 0 | 0 | $-1$ |
| $\mathbb{P}^5$ | $-1$ | $-4$ | $-1$ | 1 | 1 | 1 | 1 | 1 | 1 | 0 | 2 | $-3$ | $-3$ | $-2$ |

. (9.83)

Calculating the topological data,

$$h^\bullet_{\widetilde{\mathcal{M}}}(\widetilde{\mathcal{V}}) = (0, 72, 0, 0)\,,$$
$$h^{1,1}_{\widetilde{\mathcal{M}}} + h^{2,1}_{\widetilde{\mathcal{M}}} + h^1_{\widetilde{\mathcal{M}}}(\mathrm{End}(\widetilde{\mathcal{V}})) = 2 + 86 + 290 = 378\,. \quad (9.84)$$

we can verify that the necessary condition for a duality also holds and hence the conjecture extends to $SU(5)$ bundles as well.

## 9.3 Landscape studies

So far, we have verified the proposed general target space duality between $(0, 2)$ GLSMs only for a couple of examples. In fact, invoking a fast computer implementation, we have actually performed a large scale landscape study of this target space duality. We generated ten-thousands of candidate dual models and then computed the massless particle spectra, i.e. the number of chiral matter fields and the number of massless gauge singlets $\mathrm{D}(\mathcal{M}, \mathcal{V})$. Let us report on our findings.

## 9.3. Landscape studies

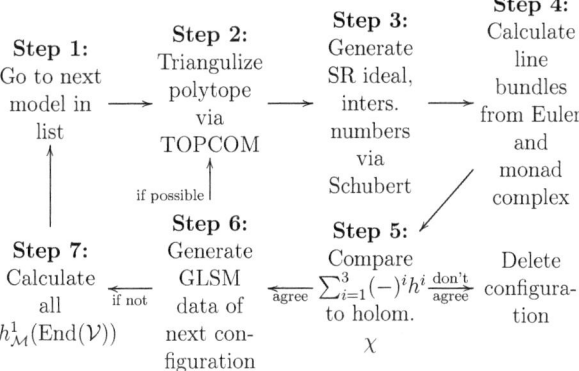

**Table 9.2.:** The scanning algorithm that we use to go through all models that has been tested for the target space duality. It starts with step 1.

### 9.3.1 The scanning algorithm

The algorithm to generate dual $(0,2)$ GLSMs enabled us to perform a scan over many different models. While one performs the duality transformation it might happen that new singularities arise and in general it may be hard to resolve them properly. For that reason, we only considered those cases where almost no new singularities appeared. Our scanning algorithm looks as described in table 9.2 starting with step 1. We ran through two different lists (mentioned in step 1). The first one contained Calabi-Yau manifolds defined via single hypersurfaces in toric varieties. We took the ambient spaces out of the list from [126] available on the website of Maximilian Kreuzer [127] and the second list contains codimension two complete intersections in weighed projective spaces which is part of the list presented in [128] and available at [129]. To resolve the ambient spaces and also to generate the set of nef partitions to obtain the codimension two Calabi-Yaus, we used PALP [63]. For the remaining steps several packages as TOPCOM [58] Schubert [59] and of course cohomCalg Koszul extension [69] along with some Mathematica routines were employed. For the interplay of TOPCOM and Schubert we use the (not published) Toric Triangulizer [64].

### 9.3.2 Hypersurfaces in toric varieties

Our first scan ran over the list of hypersurfaces in toric varieties [127] where we considered all toric varieties with 7, 8 and 9 lattice points which make altogether 1,085. Starting from this geometry, we performed all first duals to each of those models in the way described in 9.2.2 where we always introduced exactly one new hypersurface. Hence the dual models of each hypersurface Calabi-Yau are here codimension two complete intersections in toric varieties. Since already many of the duals are obtained by only performing the duality procedure once, we did not perform duals of duals as shown in 9.2.5. In figure 9.2 we displayed all models with full agreement of the chiral spectrum and the sum of complex structure, Kähler and bundle deformations, i.e. $D(\mathcal{M}, \mathcal{V})$. Some details on the full analysis are shown in table 9.3.

| Different classes | Possibly smooth models | Classes without duals | Models with matching spectrum | Models with full agreement | Computed (different) line bundle cohom. |
|---|---|---|---|---|---|
| 1,085 | 4,507 | 42 | 4,144 (100%) | 1509 (94.6%) | (1,481,539) 3,069,067 |

Table 9.3.: Some data on the landscape study: Starting point are hypersurfaces in toric varieties that are given by the polytopes with at most 9 lattice points. The percent numbers in the parentheses in column 4 and 5 only cover models where these numbers could actually be calculated. In column 5, by "full agreement" we mean that the chiral spectrum of dual models as well as the sum of complex structure, Kähler and bundle deformations agree.

### 9.3.3 CICY of two hypersurfaces

As a second scan we took a list of codimension two complete intersections in weighted projective spaces as a start, rather than just single hypersurfaces. This list can be found online at [129]. For our scan we simply ran through the first 2,780 ambient spaces and chose the 16,029 possible nef partitions as starting points. All these nef partitions correspond to topologically distinct Calabi-Yau manifolds that are complete intersections of two hypersurfaces in the corresponding weighted projective space. All dual models are codimension three complete intersections in toric varieties. In figures 9.3 and 9.4 we have displayed all the models where a full agreement of deformations and chiral spectrum was found. In table 9.4 we provide the summary of some details on the full scan.

### 9.3.4 The mismatch

While we were performing the scan over the landscape, we found that the duality holds in most the cases, but not in all of them. In two different ways it actually happened to fail i.e. either such that the chiral spectrum did not match or that

## 9.3. Landscape studies

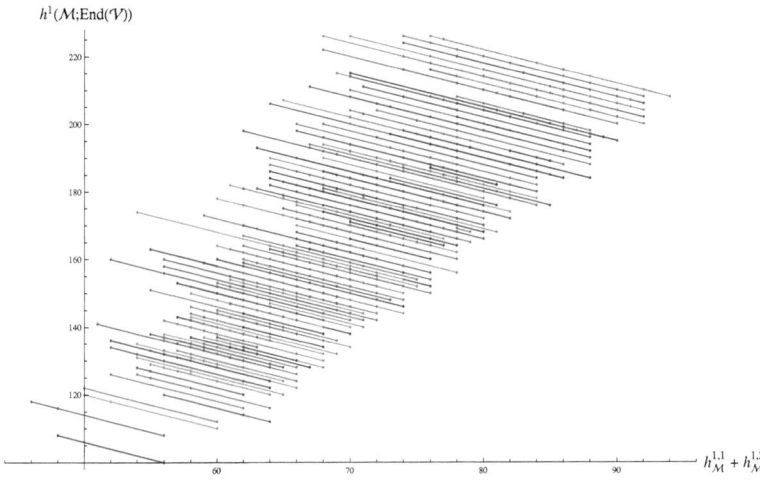

(a) Hypersurface models part 1

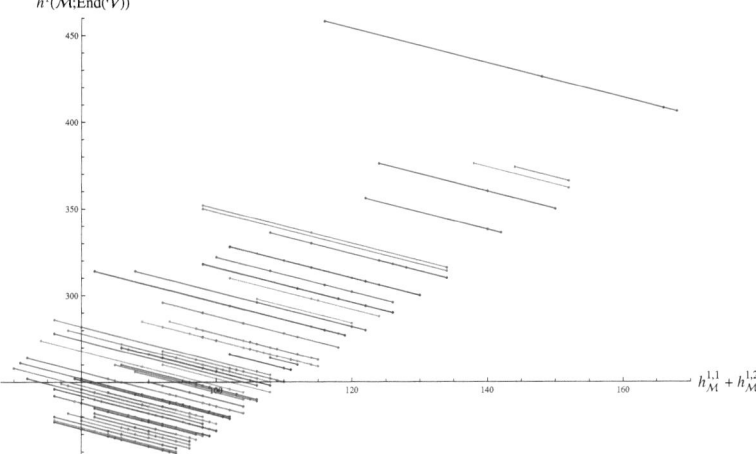

(b) Hypersurface models part 2

**Figure 9.2.:** Plot of the topological data of hypersurfaces in toric varieties and their codimension two duals with full agreement. Each line corresponds to one class of dual models. Different colored overlapping lines correspond to different classes.

194  9. Target Space Dualities in the String Landscape

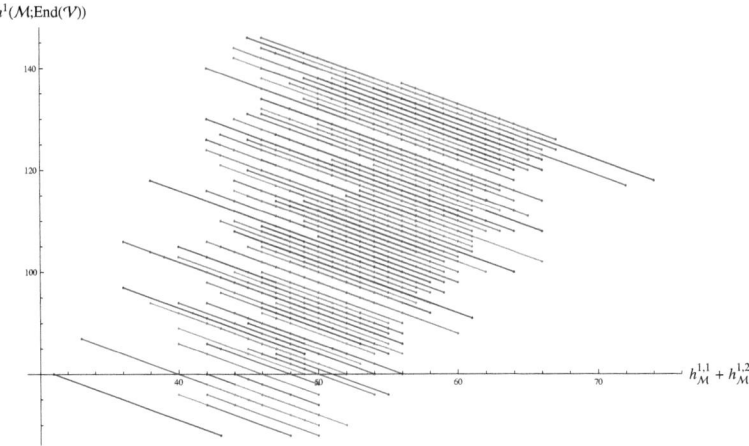

(a) Complete intersection models part 1

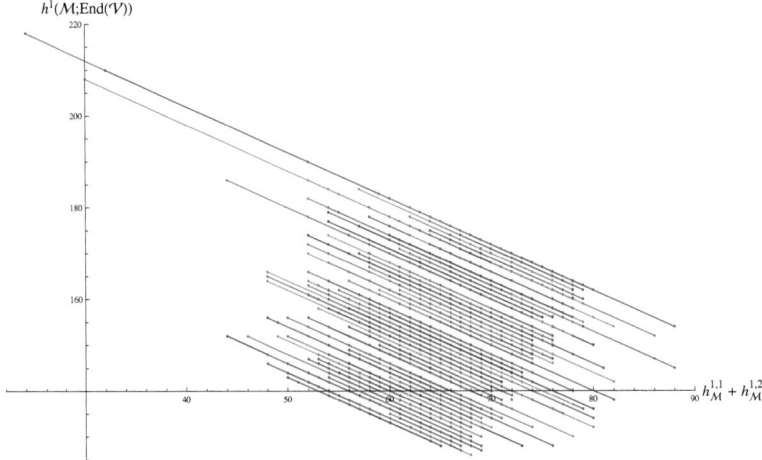

(b) Complete intersection models part 2

**Figure 9.3.:** Plot of the topological data of codimension two complete intersections in weighted projective spaces and their codimension three duals. Each line corresponds to one class of dual models. Different colored overlapping lines correspond to different classes.

## 9.3. Landscape studies

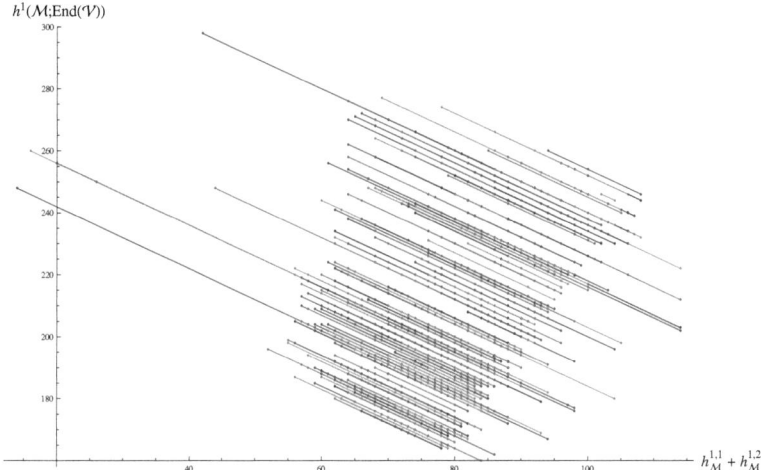

**(a)** Complete intersection models part 3

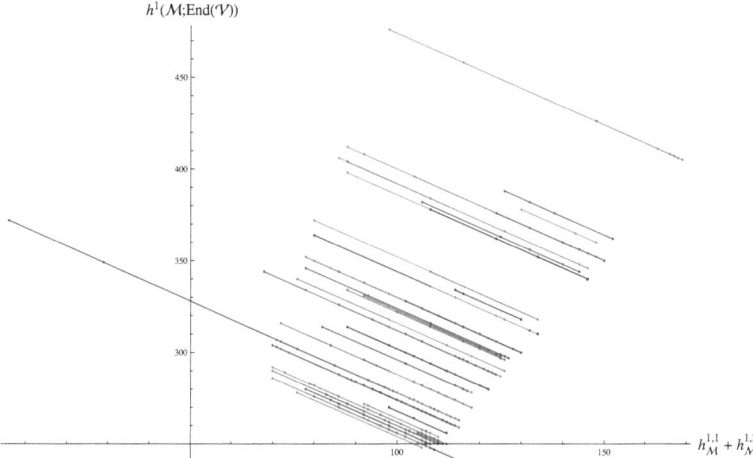

**(b)** Complete intersection models part 4

**Figure 9.4.:** Plot of the topological data of codimension two complete intersections in weighted projective spaces and their codimension three duals. Each line corresponds to one class of dual models. Different colored overlapping lines correspond to different classes.

| Different classes | Possibly smooth models | Classes without duals | Models with matching spectrum | Models with full agreement | Computed (different) line bundle cohom. |
|---|---|---|---|---|---|
| 16,961 | 79,204 | 718 | 64,332 (85 %) | 20,336 (91%) | (38,807,002) 109,228,732 |

**Table 9.4.:** Some data on landscape study. Starting point are codimension two complete intersections in weighted projective spaces.

that the sum of total deformations of the Calabi-Yau and the bundle did not match. The chiral spectrum can fail to match for the following reasons:

- The Calabi-Yau manifold is not smooth and still contains singularities[7]. In addition, the monad might not define a smooth vector bundle, but for instance merely a coherent sheaf with non-constant rank (see e.g. [115, 117] for a correct treatment of such configurations).

- During the scan we did not explicitly check whether the model we started with actually admits a phase that allows for the redefinition of the corresponding fields. So it might happen that such a phase did not exist to begin with which would forbid the exchange of specific $F$'s and $G$'s.

The mismatch of $D(\mathcal{M}, \mathcal{V}) \neq D(\widetilde{\mathcal{M}}, \widetilde{\mathcal{V}})$ could of course be traced back to the same reasons, but could also be happening since we assumed the bundle to be stable and furthermore the map $\varphi$ in (9.11) to be surjective.

**Example 9.3.1** (Failing Example). Let us present a simple example where a mismatch occurs. Consider the following configuration with standard embedding:

| $x_i$ | | | | | $\Gamma^j$ |
|---|---|---|---|---|---|
| $-1$ | 0 | 0 | 1 | 1 | $-2$ |
| 1 | 1 | 1 | 0 | 0 | $-3$ |

---

[7]The check of the holomorphic Euler characteristic for line bundles over the Calabi-Yau is only a necessary condition for smoothness.

## 9.3. Landscape studies

One possible dual model of this configuration can be obtained as

| $x_i$ | | | | | | | | $\Gamma^j$ | | $\Lambda^a$ | | | | | | $p_l$ |
|---|---|---|---|---|---|---|---|---|---|---|---|---|---|---|---|---|
| 0 | 0 | 0 | 0 | 0 | 1 | 1 | | $-1$ | $-1$ | 0 | 0 | 0 | 1 | 0 | 0 | $-1$ |
| $-1$ | 0 | 0 | 1 | 1 | 1 | 0 | 0 | $-1$ | $-1$ | $-1$ | 0 | 0 | 0 | 2 | 1 | $-2$ |
| 1 | 1 | 1 | 0 | 0 | 3 | 0 | | $-3$ | $-3$ | 1 | 1 | 1 | 0 | 0 | 0 | $-3$ |

For the initial model $(\mathcal{M}, \mathcal{V})$ we calculate the following data

$$h^\bullet_S(V) = (0, 86, 2, 0),$$
$$h^{1,1}_S + h^{2,1}_S + h^1_S(\text{End}(V)) = 2 + 86 + 184 = 272,$$

whereas for the dual model we find

$$h^\bullet_{\widetilde{S}}(\widetilde{V}) = (0, 86, 2, 0),$$
$$h^{1,1}_{\widetilde{S}} + h^{2,1}_{\widetilde{S}} + h^1_{\widetilde{S}}(\text{End}(\widetilde{V})) = 3 + 78 + 195 = 276,$$

We observe that there is a mismatch of 4 for $D(\mathcal{M}, \mathcal{V})$, but at the present state it is hard to determine the precise origin of this mismatch.

# Chapter 10

# Conclusions and Outlook

In this dissertation we have formulated and proven a mathematical theorem for the computation of line bundle cohomology. Furthermore a generalization of this algorithm to $\mathbb{Z}_n$ quotients of toric varieties was proposed. We have discovered a relation of cohomology groups of line bundles with terms in the combinatorial Batyrev formulæ which allow an identification of these cohomologies with twisted sectors of Landau-Ginzburg models. We furthermore found new explicit formulæ for the Hodge numbers of Calabi-Yau four-folds. Making use of the implementation of the algorithm, we investigated target space dualities. Here we proposed a method to construct dual models from arbitrary $(0,2)$ models. We explained the transition that the base space performs and provided evidence for the fact that this is indeed a duality of the full string model. In particular, such evidence was provided for models with various gauge groups as well as base spaces with codimension one or two. Furthermore we studied the duality for a large landscape of heterotic models.

## 10.1 Cohomology of line bundles

### Results

We have successfully developed a new algorithm to calculate the cohomology of line bundles over arbitrary toric varieties. What was already known by mathematicians is that Laurent monomials contribute to the cohomology with a certain factor that can be obtained from the cohomology of the corresponding Čech cochain complex. To avoid the lengthy computation of these cohomology groups we introduced a different sequence that is derive from the Taylor resolution of the full simplex that belongs to the Stanley-Reisner ideal, restricted to the degree of the Laurent monomial in question. Then we can obtain the Betti numbers from the reduced homology of that new complex which is in practice much faster than calculating Čech cohomology. Finally using Alexander duality one can show the relation to local cohomology with support on the irrelevant which is known to determine the desired sheaf cohomology groups. In fact one would maybe not see right away that this is indeed a faster way to compute line bundle cohomology. Here the crucial point is that we are mostly only interested in the actual dimension of these cohomologies and that this new algorithm allows us to avoid calculating the maps in the corresponding complexes for basically all relevant cases. In fact in literally millions of calculations that have been done during the last two years we did not need to calculate these maps even once. Using the conventional method via the Čech complex would require this already for the very simplest examples.

## Outlook

We also provided an efficient and conveniently accessible implementation of the algorithm along with its applications to the calculation of topological data of line bundles over subvarieties. Also the cohomology of holomorphic vector bundles can be addressed with the **cohomCalg Koszul** extension which is available online [69]. There are very many things that can be done using this algorithm and the package and we will not even try to list them all. Actually we are working on models with poly-instanton corrections right now [130] where we make use of the algorithm a lot. But in the end most of the outlook and future work is of course up to those that decide to use the algorithm for their projects.

## 10.2 Equivariant cohomology

### Results

Since the algorithm provides explicit Laurent monomials in terms of the homogeneous coordinates, we investigated the cohomology of line bundles of a space that has an additional discrete action which then splits into invariant and non-invariant parts. Here we calculated various examples using the Lefschetz index theorem to compare these equivariant cohomology groups that result from applying the discrete action to the Laurent monomials that represent it and could formulate a conjecture that goes beyond the original algorithm.

### Outlook

The above results suggest various further investigations. So far the conjecture on equivariant cohomologies only applies to the cohomology of the toric ambient space. Since in many aspects of string theory e.g. in orientifold settings we actually work with hypersurfaces it would be interesting how the representation looks in these subvarieties explicitly. Though it seems reasonable that nothing changes on invariant subvarieties it is not quite clear what happens on an arbitrary hypersurface or complete intersection of hypersurfaces, e.g. divisors of the Calabi-Yau.

## 10.3 Combinatorial cohomology

### Results

We identified the contribution of the ambient space line bundle cohomology groups to the Hodge numbers of a Calabi-Yau three-fold of codimension one and two with certain terms in the combinatorial formulæ from Batyrev and Borisov. Here we found that the terms that involve lattice points of the polytope and the dual polytope of the toric variety could be seen as contributions from cohomology of higher order. For the hypersurface case these represent twisted states in the Laudau-Ginzburg phase of the Calabi-Yau. Also a combinatorial formula for Hodge numbers of a codimension two CY four-fold were derived and presented.

## Outlook

It is quite surprising that one can actually derive a relation between the terms in the combinatorial formulæ and the line bundle cohomology degree. A nice feature of these formulæ is that they make mirror symmetry manifest. Since for the heterotic string the Hodge numbers are the moduli of the theory it would be interesting to know if a combinatorial formula also exists for the total moduli space namely the one that also takes the bundle moduli into account. In fact such a formula has been derived for the case where the holomorphic vector bundle is a deformation of the tangent bundle [108]. It would therefore be rewarding to see whether one can also identify certain terms of that formula with line bundle cohomologies. One could then maybe even go a step further and derive a more generic combinatorial formula for more general holomorphic vector bundles by deriving the combinatorial contributions from the corresponding pieces in the long exact sequences of the calculation. If that was possible one might be able to find a way to relate a holomorphic vector bundle to its $(0, 2)$ mirror.

## 10.4 Targe space dualities

### Results

We have proposed a method to construct dual models from almost any given $(0, 2)$ heterotic model that generically have the same massless spectra. This procedure should basically work with all possible structure groups $SU(3)$, $SU(4)$, $SU(5)$ for the gauge bundle and preserves certainly all anomaly cancellation conditions. For the special case that Fermi superfields become uncharged in the dual model, it was suggested to perform an additional blowup of a $\mathbb{P}^1$. Furthermore it was pointed out that in these cases the duality transformation of the base could be understood as a conifold transition.

Moreover some further aspects of target space duality were studied. For instance the successive application of the duality resulting in chains of dual models and the explicit check of the duality for models with $E_6$, $SO(10)$ and $SU(5)$ gauge group. These models correspond to Calabi-Yau manifolds endowed with a holomorphic vector bundle which cannot be obtained via a deformation of the tangent bundle.

To provide evidence for our proposal, a large number of examples were investi-

10.4. Targe space dualities

gated where the initial models were hypersurfaces in toric varieties and codimension two complete intersections in weighted projective spaces. For both types the initial configuration was the Calabi-Yau manifold equipped with a deformation of its tangent bundle carrying an $SU(3)$ structure group. A great number of models agreed in all instances and otherwise an interpretation of the mismatch of the bundle deformations and an explanation of the mismatch of the chiral spectrum was suggested.

## Outlook

A couple of aspects suggest further investigation: Since it was not checked explicitly whether the bundle of a dual configuration is indeed stable, it would be very useful to find a way or a requirement for the proposed procedure that ensures stability of the dual bundle.

We argued that deformations of the complex structure and of the bundle are unobstructed if they come from global section of line bundles on the Calabi-Yau manifold. A mathematically rigorous treatment of these obstructions was recently presented in [118] and it would be interesting whether potential target space dual models also have the same number of unobstructed deformations.

Because a large number of 83,711 models was analyzed, we are confident that a fair ratio really defines "healthy" configurations. Nevertheless, due to the fact that only necessary but not sufficient consistency checks were made in order to detect singularities of the generated spaces, a closer analysis of the specific configurations would be necessary in order to ensure that the base as well as the bundle are indeed smooth. The list of different Hodge numbers that we generated contained several yet discovered combinations of $(h^{2,1}, h^{1,1})$ and the explicit analysis of these spaces is a worthwhile thing to do.

For elliptically fibered Calabi-Yau three-folds with vector bundles defined via the spectral cover construction, it is known that a higher dimensional framework, namely F-theory on Calabi-Yau four-folds exists, in which the complex structure and bundle deformations are unified. In fact, considering a Calabi-Yau four-fold admitting two different K3-fibrations would also imply a duality between two seemingly different heterotic $(0,2)$ models. In this respect, it is an interesting question whether also in the present case of $(0,2)$ GLSMs a unified description exists, where the duality is manifest.

Having only tested single examples for the scenario where the model is not a deformation of the tangent bundle and therefore comes generically with $SU(n)$ structure for $n = 3, 4, 5$ it remains to perform a similar larger scan for such models as well. In this respect, further checks would be possible, i.e. a matching of zero modes that live in $h^1_{\mathcal{M}}(\Lambda^2 \mathcal{V})$ and $h^1_{\mathcal{M}}(\Lambda^2 \mathcal{V}^*)$. However our attempts up to now faced a couple of obstacles. Namely it is first of all not easy to solve (2.41) in general for a given base geometry. The second problem is that for a given configuration, one has to check that the bundle is not singular and that it is furthermore stable which is quite challenging. On the other hand, if one could come up with an idea to generate all stable bundles over a given base geometry systematically, it would be no problem to check (2.41) for those models. This was already done for a subset of all bundles over specific base spaces [55] and one way to prove bundle stability in an up to some point systematic way for arbitrary base spaces was suggested in [49] and [51] are a hopeful starting point to allow us to overcome this challenge.

Maybe at some point one will be able to identify the right Calabi-Yau space with the right holomorphic vector bundle resulting in a heterotic grand unified theory that describe the physics we observe. But until then:

*No GUTs no glory ...*

# Appendix A

# Anomaly Cancellation

## A.1 Anomaly cancellation for target space dualities

In this appendix, we show that the dual configuration $(\widetilde{\mathcal{M}}, \widetilde{\mathcal{V}})$ satisfy the anomaly cancellation conditions (2.41), if they were satisfied by the initial one $(\mathcal{M}, \mathcal{V})$. For this purpose, let us start with a general configuration

$$V_{N_1,...,N_{n_\Lambda}}[M_1,...,M_{n_p}] \longrightarrow \mathbb{P}_{Q_1,...,Q_d}[S_1,...,S_c].$$

that satisfies the combinatorial relations (2.41):

$$\sum_{a=1}^{n_\Lambda} N_a^{(\alpha)} = \sum_{l=1}^{n_p} M_l^{(\alpha)}, \qquad \sum_{i=1}^{d} Q_i^{(\alpha)} = \sum_{j=1}^{c} S_j^{(\alpha)}$$

$$\sum_{l=1}^{n_p} M_l^{(\alpha)} M_l^{(\beta)} - \sum_{a=1}^{n_\Lambda} N_a^{(\alpha)} N_a^{(\beta)} = \sum_{j=1}^{c} S_j^{(\alpha)} S_j^{(\beta)} - \sum_{i=1}^{d} Q_i^{(\alpha)} Q_i^{(\beta)},$$

for all $\alpha, \beta = 1,...,r$. We want to to show that this implies that the dual configuration,

$$V_{\tilde{N}_1,\tilde{N}_2,N_3,...,N_{n_\Lambda}}[M_1, M_2, ..., M_{n_p}] \longrightarrow \mathbb{P}_{Q_1,...,Q_d,B}[\tilde{S}_1, S_3, ..., S_c, \tilde{B}],$$

with charges given in table 9.1, still satisfies these relations. The new fields that changed comparing to the initial model read

| $y_1$ | $y_2$ | $\tilde{\Gamma}^1$ | $\tilde{\Gamma}^B$ | $\tilde{F}_1^1$ | $\tilde{F}_2^1$ | $\tilde{\Lambda}^1$ | $\tilde{\Lambda}^2$ | $p_1$ |
|---|---|---|---|---|---|---|---|---|
| 1 | 1 | $-1$ | $-1$ | 0 | 1 | 1 | 0 | $-1$ |
| $B$ | 0 | $-(M_1 - N_1)$ | $-(M_1 - N_2)$ | $S_1$ | $B$ | $M_1 - S_1$ | $M_1 - B$ | $-M_1$ |

.

Since $y_1$ was chosen in a way that

$$B + S_1 = ||F_1^{\ 1}|| + ||F_2^{\ 1}|| = 2M_1 - N_1 - N_2 \tag{A.1}$$

we get

$$\begin{aligned} ||\tilde{F}_2^{\ 1}|| &= \begin{pmatrix} 1 \\ 2M_1 - N_1 - N_2 - S_1 \end{pmatrix}, \\ ||\tilde{\Lambda}^2|| &= \begin{pmatrix} 1 \\ -M_1 + N_1 + N_2 + S_1 \end{pmatrix}. \end{aligned} \tag{A.2}$$

## Linear relations:

Let us refer to the $U(1)$ charges that belong to the blown up $\mathbb{P}^1$ as new $U(1)$ charges. The Calabi-Yau condition, i.e. the second equation in (2.41) for the dual model is clear for the new $U(1)$ charges. For the other $U(1)$'s it reads

$$\sum_{i=1}^{d} Q_i^{(\alpha)} + B^{(\alpha)} = \sum_{j=2}^{c} S_j^{(\alpha)} + (M_1^{(\alpha)} - N_1^{(\alpha)}) + (M_1^{(\alpha)} - N_2^{(\alpha)})$$

$$\Leftrightarrow \sum_{i=1}^{d} Q_i^{(\alpha)} + 2M_1^{(\alpha)} - N_1^{(\alpha)} - N_2^{(\alpha)} - S_1^{(\alpha)} = \sum_{j=2}^{c} S_j^{(\alpha)} + 2M_1^{(\alpha)} - N_1^{(\alpha)} - N_2^{(\alpha)}$$

$$\Leftrightarrow \sum_{i=1}^{d} Q_i^{(\alpha)} + 2M_1^{(\alpha)} - N_1^{(\alpha)} - N_2^{(\alpha)} = \sum_{j=1}^{c} S_j^{(\alpha)} + 2M_1^{(\alpha)} - N_1^{(\alpha)} - N_2^{(\alpha)}$$

$$\Leftrightarrow \sum_{i=1}^{d} Q_i^{(\alpha)} = \sum_{j=1}^{c} S_j^{(\alpha)} \qquad \square$$

The second linear relation is also satisfied:

$$\sum_{a=3}^{n_\Lambda} N_a^{(\alpha)} + (M_1 - S_1) + (-M_1 + N_1 + N_2 + S_1) = \sum_{l=1}^{n_p} M_l^{(\alpha)}$$

$$\sum_{a=1}^{n_\Lambda} N_a^{(\alpha)} = \sum_{l=1}^{n_p} M_l^{(\alpha)} \qquad \square$$

## A.1. Anomaly cancellation for target space dualities

## Quadratic relations:

Now lets have a look at the quadratic relations in (2.41). There are three different cases. The first where only the new $U(1)$ charges are involved, the second where old and new charges get mixed and the third where only the old charges are considered. Lets start with the first, which is obvious since only few changes were made:

$$1^2 + 1^2 - ((-1)^2 + (-1)^2) = 1^2 - (-1)^2 \quad \Leftrightarrow \quad 0 = 0 \quad \square$$

For the second one we find:

$$(M_1 - N_1) + (M_1 - N_2) - (B - 0) = M_1 - ((M_1 - S_1) + 0)$$
$$\Leftrightarrow 2M_1 - B = N_1 + N_2 + S_1$$
$$\Leftrightarrow 2M_1 - 2M_1 + N_1 + N_2 + S_1 = N_1 + N_2 + S_1$$
$$\Leftrightarrow N_1 + N_2 + S_1 = N_1 + N_2 + S_1 \quad \square$$

The last only involves the old $U(1)$ charges:

$$\sum_{l=1}^{n_p} M_l^{(\alpha)} M_l^{(\beta)} - \sum_{a=3}^{n_\Lambda} N_a^{(\alpha)} N_a^{(\beta)} - \tilde{N}_1^{(\alpha)} \tilde{N}_1^{(\beta)} - \tilde{N}_2^{(\alpha)} \tilde{N}_2^{(\beta)}$$
$$= \sum_{j=2}^{c} S_j^{(\alpha)} S_j^{(\beta)} + \tilde{S}_1^{(\alpha)} \tilde{S}_1^{(\beta)} + \tilde{S}_2^{(\alpha)} \tilde{S}_2^{(\beta)} - \sum_{i=1}^{d} Q_i^{(\alpha)} Q_i^{(\beta)} - B^{(\alpha)} B^{(\beta)} - 0$$
$$\Leftrightarrow \sum_{a=1}^{2} N_a^{(\alpha)} N_a^{(\beta)} - \tilde{N}_1^{(\alpha)} \tilde{N}_1^{(\beta)} - \tilde{N}_2^{(\alpha)} \tilde{N}_2^{(\beta)}$$
$$= -S_1^{(\alpha)} S_1^{(\beta)} + \tilde{S}_1^{(\alpha)} \tilde{S}_1^{(\beta)} + \tilde{S}_2^{(\alpha)} \tilde{S}_2^{(\beta)} - B^{(\alpha)} B^{(\beta)}$$
$$\Leftrightarrow \sum_{a=1}^{2} N_a^{(\alpha)} N_a^{(\beta)} - (M_1^{(\alpha)} - S_1^{(\alpha)})(M_1^{(\beta)} - S_1^{(\beta)})$$
$$\qquad -(M_1^{(\alpha)} - B^{(\alpha)})(M_1^{(\beta)} - B^{(\beta)}) \quad \text{(A.3)}$$
$$= -S_1^{(\alpha)} S_1^{(\beta)} + (M_1^{(\alpha)} - N_1^{(\alpha)})(M_1^{(\beta)} - N_1^{(\beta)})$$
$$\qquad + (M_1^{(\alpha)} - N_2^{(\alpha)})(M_1^{(\beta)} - N_2^{(\beta)}) - B^{(\alpha)} B^{(\beta)}, \quad \text{(A.4)}$$

where we used the initial quadratic relations from (2.41) in the first step. As an intermediate step, lets evaluate the third and fourth term of the right hand side of

$$-(M_1^{(\alpha)} - B^{(\alpha)})(M_1^{(\beta)} - B^{(\beta)}) =$$
$$\qquad - M_1^{(\alpha)} M_1^{(\beta)} - B^{(\alpha)} B^{(\beta)} + M_1^{(\alpha)} B^{(\beta)} + B^{(\alpha)} M_1^{(\beta)} \quad \text{(A.5)}$$

which reads

$$M_1^{(\alpha)}B^{(\beta)} = 2M_1^{(\alpha)}M_1^{(\beta)} - M_1^{(\alpha)}N_1^{(\beta)} - M_1^{(\alpha)}N_2^{(\beta)} - M_1^{(\alpha)}S_1^{(\beta)},$$
$$M_1^{(\beta)}B^{(\alpha)} = 2M_1^{(\beta)}M_1^{(\alpha)} - M_1^{(\beta)}N_1^{(\alpha)} - M_1^{(\beta)}N_2^{(\alpha)} - M_1^{(\beta)}S_1^{(\alpha)}$$

and hence it folllows

$$\begin{aligned}M_1^{(\alpha)}B^{(\beta)} + M_1^{(\beta)}B^{(\alpha)} =& M_1^{(\alpha)}M_1^{(\beta)} \\ &+ (M_1^{(\alpha)} - N_1^{(\alpha)})(M_1^{(\beta)} - N_1^{(\beta)}) - N_1^{(\alpha)}N_1^{(\beta)} \\ &+ (M_1^{(\alpha)} - N_2^{(\alpha)})(M_1^{(\beta)} - N_2^{(\beta)}) - N_2^{(\alpha)}N_2^{(\beta)} \\ &+ (M_1^{(\alpha)} - S_1^{(\alpha)})(M_1^{(\beta)} - S_1^{(\beta)}) - S_1^{(\alpha)}S_1^{(\beta)}.\end{aligned} \qquad (A.6)$$

Pluggin (A.6) back into (A.5) and (A.5) back into (A.4), we get

$$\begin{aligned}& \sum_{a=1}^{2} N_a^{(\alpha)}N_a^{(\beta)} - (M_1^{(\alpha)} - S_1^{(\alpha)})(M_1^{(\beta)} - S_1^{(\beta)}) + M_1^{(\alpha)}M_1^{(\beta)} - M_1^{(\alpha)}M_1^{(\beta)} \\ & \quad - B^{(\alpha)}B^{(\beta)} + (M_1^{(\alpha)} - N_1^{(\alpha)})(M_1^{(\beta)} - N_1^{(\beta)}) - N_1^{(\alpha)}N_1^{(\beta)} \\ & \quad + (M_1^{(\alpha)} - N_2^{(\alpha)})(M_1^{(\beta)} - N_2^{(\beta)}) - N_2^{(\alpha)}N_2^{(\beta)} \\ & \quad + (M_1^{(\alpha)} - S_1^{(\alpha)})(M_1^{(\beta)} - S_1^{(\beta)}) - S_1^{(\alpha)}S_1^{(\beta)} \\ =& \quad - S_1^{(\alpha)}S_1^{(\beta)} + (M_1^{(\alpha)} - N_1^{(\alpha)})(M_1^{(\beta)} - N_1^{(\beta)}) \\ & \quad + (M_1^{(\alpha)} - N_2^{(\alpha)})(M_1^{(\beta)} - N_2^{(\beta)}) - B^{(\alpha)}B^{(\beta)}, \\ \Leftrightarrow & \quad 0 = 0 \qquad \square.\end{aligned}$$

Since we did not assume that $M_1 = S_1$, the whole calculation is valid for both cases described in 9.2.2. In fact, it can be shown that one can exchange an arbitrary number of $G$'s with $F$'s, as long as at most one uncharged Fermi superfield appears.

# List of Figures

| | | |
|---|---|---|
| 2.1 | Dynkin diagrams of $E_8$ and its breakdown | 24 |
| 2.2 | Massless Fermions of the GLSM | 33 |
| 3.1 | Two examples of a toric variety with $r = 1$ | 41 |
| 3.2 | Toric variety with $r = 2$ | 43 |
| 3.3 | The toric diagram of $\mathbb{P}^2$ | 47 |
| 3.4 | The toric diagram of $\mathbb{P}_{122}$ | 48 |
| 3.5 | The fan of $\mathbb{P}^2$ | 49 |
| 4.1 | Pictorial representation of a long exact sequence | 58 |
| 4.2 | Chamber decomposition of $\mathcal{O}_{\mathbb{P}^2}(4)$ and $\mathcal{O}_{\mathbb{P}^2}(-7)$ | 67 |
| 4.3 | Three-dimensional simplicial complex | 74 |
| 4.4 | The toric diagram of $dP_2$ | 79 |
| 4.5 | Link of a zero-dimensional face | 84 |
| 7.1 | Split of a short complex in exact sequences | 134 |
| 9.1 | Pictorial illustration of the conifold transition | 181 |
| 9.2 | Landscape plot: Hypersurfaces in toric varieties | 193 |
| 9.3 | Landscape plot: Codimension two subvarieties (1/2) | 194 |
| 9.4 | Landscape plot: Codimension two subvarieties (2/2) | 195 |

# List of Tables

| | | |
|---|---|---|
| 2.1 | Matter zero modes in representations of the GUT group | 25 |
| 4.1 | Toric data for the del Pezzo-2 surface | 82 |
| 4.2 | Sequences corresponding to the two contributing LMs of $\mathbb{P}^2$ | 92 |
| 4.3 | Toric data for the del Pezzo-1 surface | 93 |
| 4.4 | Sequences corresponding to the contributing LMs of $dP_1$ | 93 |
| 4.5 | Toric data for the del Pezzo-3 surface. | 96 |
| 4.6 | Sequences corresponding to a particular LM of $dP_3$ | 96 |
| 5.1 | Toric data for the non-generic $\widetilde{dP_5}$. | 105 |
| 5.2 | Toric data for the del Pezzo-1 surface | 112 |
| 5.3 | Toric data for the del Pezzo-3 surface. | 114 |
| 7.1 | General Hodge diamond of a three-fold | 131 |
| 7.2 | General Hodge diamond with Poincare duality applied | 131 |
| 7.3 | General Hodge diamond of a Calabi-Yau three-fold | 132 |
| 7.4 | Toric data of the toric blowup of $\mathbb{P}^4_{11222}$ | 136 |
| 7.5 | Toric data for $\mathbb{P}^5_{111122}[4,4]$ | 140 |
| 7.6 | Sequences to determine $\Lambda^2 A, \Lambda^2 B$ and $\Lambda^2 C$. | 146 |
| 8.1 | Lattice polytope $\Delta^\circ$ for $\mathbb{P}^5_{112233}$ | 158 |
| 8.2 | (Dual) Cayley polytope for a nef partition of $\mathbb{P}^5_{112233}$ | 159 |
| 8.3 | Line bundle cohomology contribution to $h^{1,1}(\mathcal{S})$ | 163 |
| 8.4 | Line bundle cohomology contribution to $h^{2,1}(\mathcal{S})$ | 163 |
| 9.1 | Charges and degrees of the dual model | 176 |
| 9.2 | The scanning algorithm | 191 |
| 9.3 | Some data on the landscape study: Codimension one | 192 |
| 9.4 | Some data on the landscape study: Codimension two | 196 |

# Bibliography

[1] S. L. Glashow, "Partial Symmetries of Weak Interactions," *Nucl. Phys.* **22** (1961) 579–588.

[2] S. Weinberg, "A Model of Leptons," *Phys. Rev. Lett.* **19** (1967) 1264–1266.

[3] A. Salam and J. C. Ward, "Electromagnetic and weak interactions," *Phys. Lett.* **13** (1964) 168–171.

[4] S. Weinberg, "The cosmological constant problem," *Rev. Mod. Phys.* **61** (1989) 1–23.

[5] H. Georgi and S. L. Glashow, "Unity of All Elementary Particle Forces," *Phys. Rev. Lett.* **32** (1974) 438–441.

[6] H. Fritzsch and P. Minkowski, "Unified Interactions of Leptons and Hadrons," *Ann. Phys.* **93** (1975) 193–266.

[7] H. P. Nilles, "Supersymmetry, Supergravity and Particle Physics," *Phys. Rept.* **110** (1984) 1–162.

[8] J. Wess and J. Bagger, *Supersymmetry and supergravity*. Princeton series in physics. Princeton University Press, 1992.

[9] F. Quevedo, S. Krippendorf, and O. Schlotterer, "Cambridge Lectures on Supersymmetry and Extra Dimensions," 1011.1491.

[10] S. R. Coleman and J. Mandula, "ALL POSSIBLE SYMMETRIES OF THE S MATRIX," *Phys. Rev.* **159** (1967) 1251–1256.

[11] S. Dimopoulos and H. Georgi, "Softly Broken Supersymmetry and SU(5)," *Nucl. Phys.* **B193** (1981) 150.

[12] S. Dimopoulos, S. Raby, and F. Wilczek, "Supersymmetry and the Scale of Unification," *Phys. Rev.* **D24** (1981) 1681–1683.

[13] M. Dine, W. Fischler, and M. Srednicki, "A Simple Solution to the Strong CP Problem with a Harmless Axion," *Phys. Lett.* **B104** (1981) 199.

[14] H. P. Nilles, M. Srednicki, and D. Wyler, "Weak Interaction Breakdown Induced by Supergravity," *Phys. Lett.* **B120** (1983) 346.

[15] J. M. Frere, D. R. T. Jones, and S. Raby, "Fermion Masses and Induction of the Weak Scale by Supergravity," *Nucl. Phys.* **B222** (1983) 11.

[16] J. P. Derendinger and C. A. Savoy, "Quantum Effects and SU(2) x U(1) Breaking in Supergravity Gauge Theories," *Nucl. Phys.* **B237** (1984) 307.

[17] M. B. Green and J. H. Schwarz, "Anomaly Cancellation in Supersymmetric D=10 Gauge Theory and Superstring Theory," *Phys. Lett.* **B149** (1984) 117–122.

[18] F. Gliozzi, J. Scherk, and D. I. Olive, "Supersymmetry, Supergravity Theories and the Dual Spinor Model," *Nucl. Phys.* **B122** (1977) 253–290.

[19] E. Witten, "String theory dynamics in various dimensions," *Nucl. Phys.* **B443** (1995) 85–126, hep-th/9503124.

[20] R. Donagi, B. A. Ovrut, T. Pantev, and D. Waldram, "Standard models from heterotic M theory," *Adv.Theor.Math.Phys.* **5** (2002) 93–137, hep-th/9912208.

[21] V. Bouchard and R. Donagi, "An SU(5) heterotic standard model," *Phys.Lett.* **B633** (2006) 783–791, hep-th/0512149.

[22] V. Braun, Y.-H. He, B. A. Ovrut, and T. Pantev, "A Heterotic standard model," *Phys.Lett.* **B618** (2005) 252–258, hep-th/0501070.

[23] V. Braun, Y.-H. He, B. A. Ovrut, and T. Pantev, "A Standard model from the E(8) × E(8) heterotic superstring," *JHEP* **0506** (2005) 039, hep-th/0502155.

[24] A. Bak, V. Bouchard, and R. Donagi, "Exploring a new peak in the heterotic landscape," *JHEP* **1006** (2010) 108, 0811.1242.

[25] L. B. Anderson, Y.-H. He, and A. Lukas, "Monad Bundles in Heterotic String Compactifications," *JHEP* **0807** (2008) 104, 0805.2875.

[26] L. B. Anderson, J. Gray, A. Lukas, and E. Palti, "Two Hundred Heterotic Standard Models on Smooth Calabi-Yau Threefolds," *Phys.Rev.* **D84** (2011) 106005, 1106.4804. 19 pages. References Added.

[27] L. B. Anderson, J. Gray, A. Lukas, and E. Palti, "Heterotic Line Bundle Standard Models," 1202.1757.

[28] R. Blumenhagen, B. Kors, D. Lust, and S. Stieberger, "Four-dimensional String Compactifications with D-Branes, Orientifolds and Fluxes," *Phys. Rept.* **445** (2007) 1–193, hep-th/0610327.

[29] T. Dijkstra, L. Huiszoon, and A. Schellekens, "Supersymmetric standard model spectra from RCFT orientifolds," *Nucl.Phys.* **B710** (2005) 3–57, hep-th/0411129.

[30] T. Dijkstra, L. Huiszoon, and A. Schellekens, "Chiral supersymmetric standard model spectra from orientifolds of Gepner models," *Phys.Lett.* **B609** (2005) 408–417, hep-th/0403196.

[31] W. Buchmuller, K. Hamaguchi, O. Lebedev, and M. Ratz, "Supersymmetric standard model from the heterotic string," *Phys.Rev.Lett.* **96** (2006) 121602, hep-ph/0511035.

[32] W. Buchmuller, K. Hamaguchi, O. Lebedev, and M. Ratz, "Supersymmetric Standard Model from the Heterotic String (II)," *Nucl.Phys.* **B785** (2007) 149–209, hep-th/0606187.

[33] W. Buchmuller, K. Hamaguchi, O. Lebedev, S. Ramos-Sanchez, and M. Ratz, "Seesaw neutrinos from the heterotic string," *Phys.Rev.Lett.* **99** (2007) 021601, hep-ph/0703078.

[34] O. Lebedev, H. P. Nilles, S. Raby, S. Ramos-Sanchez, M. Ratz, *et al.*, "The Heterotic Road to the MSSM with R parity," *Phys.Rev.* **D77** (2008) 046013, 0708.2691.

[35] O. Lebedev, H. P. Nilles, S. Ramos-Sanchez, M. Ratz, and P. K. Vaudrevange, "Heterotic mini-landscape. (II). Completing the search for MSSM vacua in a Z(6) orbifold," *Phys.Lett.* **B668** (2008) 331–335, 0807.4384.

[36] J. E. Kim, J.-H. Kim, and B. Kyae, "Superstring standard model from Z(12-I) orbifold compactification with and without exotics, and effective R-parity," *JHEP* **0706** (2007) 034, hep-ph/0702278.

[37] M. Blaszczyk, S. Nibbelink Groot, M. Ratz, F. Ruehle, M. Trapletti, *et al.*, "A Z2xZ2 standard model," *Phys.Lett.* **B683** (2010) 340–348, 0911.4905.

[38] A. E. Faraggi, D. V. Nanopoulos, and K.-j. Yuan, "A Standard Like Model in the 4D Free Fermionic String Formulation," *Nucl.Phys.* **B335** (1990) 347.

[39] A. E. Faraggi, "A New standard - like model in the four-dimensional free fermionic string formulation," *Phys.Lett.* **B278** (1992) 131–139.

[40] G. B. Cleaver, A. E. Faraggi, and S. Nooij, "NAHE based string models with SU(4) x SU(2) x U(1) SO(10) subgroup," *Nucl.Phys.* **B672** (2003) 64–86, hep-ph/0301037.

[41] A. E. Faraggi, E. Manno, and C. Timirgaziu, "Minimal Standard Heterotic String Models," *Eur.Phys.J.* **C50** (2007) 701–710, hep-th/0610118.

[42] C. Closset, "Toric geometry and local Calabi-Yau varieties: An Introduction to toric geometry (for physicists)," 0901.3695.

[43] H. Skarke, "String dualities and toric geometry: An Introduction," *Chaos Solitons Fractals* (1998) hep-th/9806059.

[44] T. Hübsch, *Calabi-Yau manifolds: a bestiary for physicists*. World Scientific, 1994.

[45] W. Fulton, *Introduction to Toric Varieties*. Princeton University Press, 1993.

[46] D. Cox, J. Little, and H. Schenck, *Toric Varieties*. Graduate studies in mathematics. American Mathematical Society, 2011.

[47] J. Gray, Y.-H. He, V. Jejjala, and B. D. Nelson, "Exploring the vacuum geometry of N=1 gauge theories," *Nucl.Phys.* **B750** (2006) 1–27, hep-th/0604208.

[48] L. B. Anderson, Y.-H. He, and A. Lukas, "Heterotic Compactification, An Algorithmic Approach," *JHEP* **0707** (2007) 049, hep-th/0702210.

[49] L. B. Anderson, "Heterotic and M-theory Compactifications for String Phenomenology," 0808.3621.

[50] M. Gabella, Y.-H. He, and A. Lukas, "An Abundance of Heterotic Vacua," *JHEP* **0812** (2008) 027, 0808.2142.

[51] L. B. Anderson, J. Gray, A. Lukas, and B. Ovrut, "The Edge Of Supersymmetry: Stability Walls in Heterotic Theory," *Phys.Lett.* **B677** (2009) 190–194, 0903.5088.

[52] L. B. Anderson, J. Gray, D. Grayson, Y.-H. He, and A. Lukas, "Yukawa Couplings in Heterotic Compactification," *Commun.Math.Phys.* **297** (2010) 95–127, 0904.2186.

[53] Y.-H. He, S.-J. Lee, and A. Lukas, "Heterotic Models from Vector Bundles on Toric Calabi-Yau Manifolds," *JHEP* **1005** (2010) 071, 0911.0865.

[54] L. B. Anderson, J. Gray, A. Lukas, and B. Ovrut, "Stability Walls in Heterotic Theories," *JHEP* **0909** (2009) 026, 0905.1748.

[55] L. B. Anderson, J. Gray, Y.-H. He, and A. Lukas, "Exploring Positive Monad Bundles And A New Heterotic Standard Model," *JHEP* **1002** (2010) 054, 0911.1569.

[56] L. B. Anderson, J. Gray, and B. Ovrut, "Yukawa Textures From Heterotic Stability Walls," *JHEP* **1005** (2010) 086, 1001.2317.

[57] L. B. Anderson, V. Braun, R. L. Karp, and B. A. Ovrut, "Numerical Hermitian Yang-Mills Connections and Vector Bundle Stability in Heterotic Theories," *JHEP* **1006** (2010) 107, 1004.4399.

[58] J. Rambau, "TOPCOM: Triangulations of point configurations and oriented matroids," in *Mathematical Software—ICMS 2002*, A. M. Cohen, X.-S. Gao, and N. Takayama, eds., pp. 330–340. World Scientific, 2002.

[59] S. Katz, S. A. Stromme, and J.-M. Økland, "Schubert." Package for intersection theory and enumerative geometry, http://folk.uib.no/nmasr/schubert/, 1992-2006.

[60] D. R. Grayson and M. E. Stillman, "Macaulay2, a software system for research in algebraic geometry." http://www.math.uiuc.edu/Macaulay2/.

[61] W. Stein *et al.*, *Sage Mathematics Software*. The Sage Development Team, 2010. http://www.sagemath.org.

[62] W. Decker, G.-M. Greuel, G. Pfister, and H. Schönemann, "SINGULAR 3-1-3 — A computer algebra system for polynomial computations,". http://www.singular.uni-kl.de.

[63] M. Kreuzer and H. Skarke, "PALP: A Package for analyzing lattice polytopes with applications to toric geometry," *Comput.Phys.Commun.* **157** (2004) 87–106, math/0204356.

[64] B. Jurke, "The Toric Triangulizer." Unpublished C++ wrapper for TOPCOM, Maple/SCHUBERT and associated Mathematica scripts., 2009.

[65] J. Gray, Y.-H. He, A. Ilderton, and A. Lukas, "STRINGVACUA: A Mathematica Package for Studying Vacuum Configurations in String Phenomenology," *Comput.Phys.Commun.* **180** (2009) 107–119, 0801.1508.

[66] R. Blumenhagen, B. Jurke, T. Rahn, and H. Roschy, "Cohomology of Line Bundles: A Computational Algorithm," *J. Math. Phys.* **51** (2010) 103525, 1003.5217.

[67] T. Rahn and H. Roschy, "Cohomology of Line Bundles: Proof of the Algorithm," *J. Math. Phys.* **51** (2010) 103520, 1006.2392.

[68] S.-Y. Jow, "Cohomology of toric line bundles via simplicial Alexander duality," *Journal of Mathematical Physics* **52** (Mar., 2011) 033506, 1006.0780.

[69] "cohomCalg Koszul extension." Download link, 2010. High-performance line bundle cohomology computation based on [66], http://wwwth.mppmu.mpg.de/members/blumenha/cohomcalg/.

[70] R. Blumenhagen, B. Jurke, T. Rahn, and H. Roschy, "Cohomology of Line Bundles: Applications," *J. Math. Phys.* **53** (2012) 012302, 1010.3717.

[71] R. Blumenhagen, B. Jurke, and T. Rahn, "Computational Tools for Cohomology of Toric Varieties," *Adv. High Energy Phys.* **2011** (2011) 152749, 1104.1187.

[72] L. J. Dixon, J. A. Harvey, C. Vafa, and E. Witten, "Strings on Orbifolds. 2," *Nucl. Phys.* **B274** (1986) 285–314.

[73] L. J. Dixon, J. A. Harvey, C. Vafa, and E. Witten, "Strings on Orbifolds," *Nucl. Phys.* **B261** (1985) 678–686.

[74] Y. Katsuki *et al.*, "Z(N) ORBIFOLD MODELS," *Nucl. Phys.* **B341** (1990) 611–640.

[75] T. Kobayashi and N. Ohtsubo, "Geometrical aspects of Z(N) orbifold phenomenology," *Int. J. Mod. Phys.* **A9** (1994) 87–126.

[76] V. V. Batyrev, "Dual polyhedra and mirror symmetry for Calabi-Yau hypersurfaces in toric varieties," *J. Alg. Geom.* **3** (1994) 493–545.

[77] R. Blumenhagen and T. Rahn, "Landscape Study of Target Space Duality of (0,2) Heterotic String Models," *JHEP* **09** (2011) 098, 1106.4998.

[78] T. Rahn, "Target Space Dualities of Heterotic Grand Unified Theories," 1111.0491.

[79] V. V. Batyrev and L. A. Borisov, "On calabi-yau complete intersections in toric varieties," alg-geom/9412017.

[80] D. J. Gross, J. A. Harvey, E. J. Martinec, and R. Rohm, "Heterotic String Theory. 1. The Free Heterotic String," *Nucl. Phys.* **B256** (1985) 253.

[81] D. J. Gross, J. A. Harvey, E. J. Martinec, and R. Rohm, "The Heterotic String," *Phys. Rev. Lett.* **54** (1985) 502–505.

[82] S.-T. Yau, "Calabi's Conjecture and some new results in algebraic geometry," *Proc. Nat. Acad. Sci.* **74** (1977) 1798–1799.

[83] S. K. Donaldson, "Anti self-dual yang-mills connections over complex algebraic surfaces and stable vector bundles," *Proceedings of the London Mathematical Society* **s3-50** (1985), no. 1, 1–26.

[84] K. Uhlenbeck and S. T. Yau, "On the existence of hermitian-yang-mills connections in stable vector bundles," *Communications on Pure and Applied Mathematics* **39** (1986), no. S1, S257–S293.

[85] E. Witten, "New Issues in Manifolds of SU(3) Holonomy," *Nucl. Phys.* **B268** (1986) 79.

[86] R. Donagi, A. Lukas, B. A. Ovrut, and D. Waldram, "Holomorphic vector bundles and non-perturbative vacua in M-theory," *JHEP* **06** (1999) 034, hep-th/9901009.

[87] R. Donagi, A. Lukas, B. A. Ovrut, and D. Waldram, "Non-perturbative vacua and particle physics in M-theory," *JHEP* **05** (1999) 018, hep-th/9811168.

[88] E. Witten, "Phases of N = 2 theories in two dimensions," *Nucl. Phys.* **B403** (1993) 159–222, hep-th/9301042.

[89] P. S. Aspinwall, B. R. Greene, and D. R. Morrison, "Calabi-Yau moduli space, mirror manifolds and spacetime topology change in string theory," *Nucl. Phys.* **B416** (1994) 414–480, hep-th/9309097.

[90] J. Distler, "Notes on (0,2) superconformal field theories," hep-th/9502012.

[91] J. McOrist, "The Revival of (0,2) Linear Sigma Models," *Int.J.Mod.Phys.* **A26** (2011) 1–41, 1010.4667.

[92] J. Distler and S. Kachru, "(0,2) Landau-Ginzburg theory," *Nucl. Phys.* **B413** (1994) 213–243, hep-th/9309110.

[93] J. Distler and S. Kachru, "Singlet couplings and (0,2) models," *Nucl. Phys.* **B430** (1994) 13–30, hep-th/9406090.

[94] J. Distler and S. Kachru, "Duality of (0,2) string vacua," *Nucl.Phys.* **B442** (1995) 64–74, hep-th/9501111.

[95] T.-M. Chiang, J. Distler, and B. R. Greene, "Some features of (0,2) moduli space," *Nucl.Phys.* **B496** (1997) 590–616, hep-th/9702030.

[96] A. Adams, A. Basu, and S. Sethi, "(0,2) duality," *Adv. Theor. Math. Phys.* **7** (2004) 865–950, hep-th/0309226.

[97] M. Larfors, D. Lust, and D. Tsimpis, "Flux compactification on smooth, compact three-dimensional toric varieties," *JHEP* **1007** (2010) 073, 1005.2194.

[98] R. Blumenhagen, V. Braun, T. W. Grimm, and T. Weigand, "GUTs in Type IIB Orientifold Compactifications," *Nucl. Phys.* **B815** (2009) 1–94, 0811.2936.

[99] R. Blumenhagen, A. Collinucci, and B. Jurke, "On Instanton Effects in F-theory," *JHEP* **08** (2010) 079, 1002.1894.

[100] C. Weibel, *An introduction to homological algebra.* Cambridge studies in advanced mathematics. Cambridge University Press, 1995.

[101] D. Eisenbud, M. Mustata, and M. Stillman, "Cohomology on Toric Varieties and Local Cohomology with Monomial Supports," *ArXiv Mathematics e-prints* (Jan., 2000) arXiv:math/0001159.

[102] P. Shanahan, *The Atiyah-Singer Index Theorem.* Springer, Berlin, 1978.

[103] M. Cvetic, I. Garcia-Etxebarria, and J. Halverson, "On the computation of non-perturbative effective potentials in the string theory landscape – IIB/F-theory perspective," 1009.5386.

[104] E. Calabi, "On Kähler manifolds with vanishing canonical class," in *Algebraic geometry and topology. A symposium in honor of S. Lefschetz*, pp. 78–89. Princeton University Press, Princeton, N. J., 1957.

[105] P. Griffiths and J. Harris, *Principles of algebraic geometry*. Wiley New York, 1994.

[106] L. B. Anderson, "Heterotic and M-theory Compactifications for String Phenomenology," *ArXiv e-prints* (Aug., 2008) 0808.3621.

[107] V. V. Batyrev and L. A. Borisov, "Mirror duality and string-theoretic Hodge numbers," alg-geom/9509009.

[108] M. Kreuzer, J. McOrist, I. V. Melnikov, and M. R. Plesser, "(0,2) Deformations of Linear Sigma Models," *JHEP* **07** (2011) 044, 1001.2104.

[109] I. V. Melnikov and M. R. Plesser, "A (0,2) Mirror Map," *JHEP* **02** (2011) 001, 1003.1303.

[110] K. Hori, S. Katz, A. Klemm, R. Pandharipande, R. Thomas, C. Vafa, R. Vakil, and E. Zaslow, *Mirror Symmetry*, vol. 1 of *Clay Mathematics Monograph*, ch. Toric Geometry for String Theory. AMS Clay Mathematics Institute, 2003.

[111] C. F. Doran and A. Y. Novoseltsev, "Closed form expressions for Hodge numbers of complete intersection Calabi-Yau threefolds in toric varieties," *ArXiv e-prints* (July, 2009) 0907.2701.

[112] V. Batyrev and B. Nill, "Combinatorial aspects of mirror symmetry," math/0703456.

[113] B. Nill and J. Schepers, "Gorenstein polytopes and their stringy E-functions," 1005.5158.

[114] R. Birkner, "Polyhedra: a package for computations with convex polyhedral objects." Download link, 2009.

[115] J. Distler, B. R. Greene, and D. R. Morrison, "Resolving singularities in (0,2) models," *Nucl.Phys.* **B481** (1996) 289–312, hep-th/9605222.

[116] R. Blumenhagen, "Target space duality for (0,2) compactifications," *Nucl.Phys.* **B513** (1998) 573–590, hep-th/9707198.

[117] R. Blumenhagen, "(0,2) Target space duality, CICYs and reflexive sheaves," *Nucl.Phys.* **B514** (1998) 688–704, hep-th/9710021.

[118] L. B. Anderson, J. Gray, A. Lukas, and B. Ovrut, "Stabilizing the Complex Structure in Heterotic Calabi-Yau Vacua," *JHEP* **1102** (2011) 088, 1010.0255.

[119] J. Distler and B. R. Greene, "Aspects of (2,0) String Compactifications," *Nucl. Phys.* **B304** (1988) 1.

[120] E. Silverstein and E. Witten, "Criteria for conformal invariance of (0,2) models," *Nucl. Phys.* **B444** (1995) 161–190, hep-th/9503212.

[121] A. Basu and S. Sethi, "World-sheet stability of (0,2) linear sigma models," *Phys. Rev.* **D68** (2003) 025003, hep-th/0303066.

[122] C. Beasley and E. Witten, "Residues and world-sheet instantons," *JHEP* **10** (2003) 065, hep-th/0304115.

[123] P. S. Aspinwall, I. V. Melnikov, and M. Plesser, "(0,2) Elephants," *JHEP* **1201** (2012) 060, 1008.2156.

[124] P. S. Aspinwall and M. R. Plesser, "Elusive Worldsheet Instantons in Heterotic String Compactifications," 1106.2998.

[125] P. Candelas, P. S. Green, and T. Hubsch, "Rolling Among Calabi-Yau Vacua," *Nucl.Phys.* **B330** (1990) 49.

[126] M. Kreuzer and H. Skarke, "Complete classification of reflexive polyhedra in four-dimensions," *Adv.Theor.Math.Phys.* **4** (2002) 1209–1230, hep-th/0002240.

[127] M. Kreuzer and H. Skarke, "http://tph16.tuwien.ac.at/ kreuzer/CY/."

[128] A. Klemm, M. Kreuzer, E. Riegler, and E. Scheidegger, "Topological string amplitudes, complete intersection Calabi-Yau spaces and threshold corrections," *JHEP* **0505** (2005) 023, hep-th/0410018.

[129] A. Klemm, M. Kreuzer, E. Riegler, and E. Scheidegger, "http://hep.itp.tuwien.ac.at/ kreuzer/CY/hep-th/0410018.html."

[130] R. Blumenhagen, X. Gao, T. Rahn, and P. Shukla, "Moduli Stabilization and Inflation from Poly-Instantons in Type IIB Orientifolds." work in progress.

# i want morebooks!

Buy your books fast and straightforward online - at one of world's fastest growing online book stores! Environmentally sound due to Print-on-Demand technologies.

Buy your books online at
## www.get-morebooks.com

Kaufen Sie Ihre Bücher schnell und unkompliziert online – auf einer der am schnellsten wachsenden Buchhandelsplattformen weltweit! Dank Print-On-Demand umwelt- und ressourcenschonend produziert.

Bücher schneller online kaufen
## www.morebooks.de

VDM Verlagsservicegesellschaft mbH
Heinrich-Böcking-Str. 6-8          Telefon: +49 681 3720 174          info@vdm-vsg.de
D - 66121 Saarbrücken              Telefax: +49 681 3720 1749         www.vdm-vsg.de

Printed by Books on Demand GmbH, Norderstedt / Germany